Miscellaneous
173
605. THE PET HORSE.

Meadville, Pa., New York, N.Y., Portland, Oregon, London, Eng., Sydney, Aus.

PHOTOGRAPHED AND PUBLISHED BY H. H. BENNETT, Kilbourn City, Wis.

WAYSIDE GEMS.
379 Jack Frost's Aquarium.

3257
SCENES IN ALL PARTS OF THE WORLD.
WILLIAM H. RAU, Publisher.
PHILADELPHIA, PA.
3257 Belgium The Giraffe Antwerp Zoo Copyrighted 1904 by William H. Rau

SCHREIBER & SONS, PHILADELPHIA

TUDIES FROM NATURE

Copyright 1898, by B. W. Kilburn.
12401. Tomb of Khalifs, Cairo, Egypt.

WATKINS' NEW S
Of Pacific Coast Views. 427 Montgomery Str

WANDERINGS AMONG THE WONDERS AND BEAUTIES OF
Wisconsin Scenery.
WAYSIDE GEMS.
rost's Aquarium.

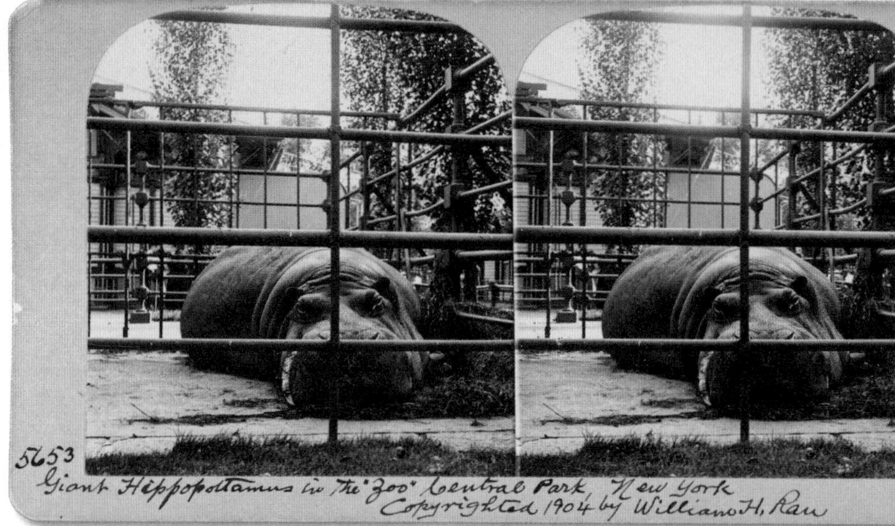
5653
Giant Hippopotamus in the "Zoo" Central Park New York
Copyrighted 1904 by William H. Rau

ANIMAL STUDIES FROM NATURE.

INSTANTANEOUS PORTRAITS OF
ANIMALS A SPECIALITY.

SCHREIBER & SONS' PH

N WASHINGTON D. C.

WHY WE PHOTOGRAPH ANIMALS

WHY WE PHOTOGRAPH ANIMALS

HUW LEWIS-JONES

with 278 illustrations

Contents

Introduction

Tel-El-Kebir

Ch. Clint Chief

Flair of Bowland

Ch. Portencross Prince

Runlee Peter Pan

Ch. Basford Revival Replica

'I like dogs. They're sympathetic. They're nice. They don't ask for prints.'

Elliott Erwitt

IT IS SOMETIMES said that photography is the most widespread and important medium in contemporary visual culture, and yet it is also one that is now cursed by its abundance. Forgive me then, if I begin this story with a billion dogs ...

Dogs have been with us for thousands of years, arguably the first animal welcomed to a fire's side by our early ancestors. They could be trained to hunt and help protect the tribe, as well as being fun to have around. In the comparatively short history of photography – little more than 180 years – it could also be argued that dogs are the world's most photographed animal. It has been estimated that over ninety-five million photos and videos are uploaded to Instagram every twenty-four hours. That includes more than three million dog photos every day in the UK alone! If you consider all the other social media platforms, we're talking billions of posted images, and even this is just a fraction of the photos created and carried around in our pockets. It's a paradox though, as curator Jeffrey Fraenkel has written, 'that photography's ubiquity, and the ease with which it can be viewed, swiped, and deleted, has made it considerably harder to see.'

'Nature is painting for us, day after day,' wrote John Ruskin, 'pictures of infinite beauty, if only we have the eyes to see them.' So, are we really paying attention? Photography can bear witness to the world's wonders and also to its horrors. Our hearts expand in the presence of images of power and beauty and grace, but some truths these days are too hard to bear. Photography is a continuing technological marvel but we've been trashing the planet and using animals at every step in our history.

We could try to begin a collection of this kind with the *first* animal photograph. We will show many rare images here, but it is an impossible task to find the first with any certainty as so much from the earliest years of photographic invention is lost to us. Maybe it was a horse standing tethered to a carriage on the side of a Parisian boulevard? An important animal the horse, an animal that enabled the beginnings of the world we know, an animal of war and transport and trade. Or perhaps a parrot, collected as an exotic prop? Or an insect's wing, fragile and diaphanous? Early experiments in photographing animals do survive, but the story of the animal *in* photography might rightly begin elsewhere in time. So, what about a different kind of pet, a royal hippo? Or a stuffed heron and a fake stag? And what about a galloping horse, or a Hollywood cougar? A gorilla selfie, even?

'Champion Dogs', cigarette cards, issued by John Sinclair Ltd, 1938

–

This collectible set of cards featured winners at Crufts dog show. The very first show in 1891 attracted over two thousand entries and photography played an integral part in showcasing and popularizing new breeds.

All these shots and more are here. They were created for a host of reasons and they enjoy varied afterlives. Finding our way among a deluge of photographic images, we will head in many different directions. Some pictures here are famous, others are deliberately esoteric. Some have been created by well-known photographers, others by folks whose identities will forever remain unknown. The images are curated to raise questions, gathered here as an extended visual essay; joined together for the first time, they help us to interrogate the development of the medium and question our evolving understanding of animals. As with most photos, their meanings can change each time they are seen.

Antoine Claudet, hand-coloured daguerreotype, London, c. 1856
–
Claudet was elected a fellow of the Royal Society in 1853 and appointed 'photographer-in-ordinary' to Queen Victoria. His showroom and studio at 107 Regent Street, where this image was made, was known as the Temple of Photography. A month after his death in 1867 the studio burned down and most of his images were lost.

* * *

Humans have been making pictures of the animals they live alongside since the beginnings of what we might understand as art. Whether painted on a cave wall or carved in bone, the imaged animal – the *imagined* animal – is story as much as flesh, fur, tooth and claw. In a history of culture many thousands of years in the making, photography's arrival is but a heartbeat ago. Yet given its short history, its presence in our lives is now almost unfathomable.

When I give talks about photography, I sometimes ask people in the audience to raise their hand if they *don't* own a camera: very few these days, possibly none at all. Everyone has a phone. And animals are there, everywhere. Scroll through images on your device and I guarantee you'll find an animal, even if you think you've never photographed one. A pet? A bird in the garden, or on your holiday at the beach? Happy memories and forgettable memes. Billions of dogs, as I said, but also cats with hats, trips to the zoo, even penguins if you're lucky.

People want to see animals, to photograph them and to share images of them. The desire to keep looking at animals is strong, even as they are disappearing in the wild. A recent study revealed that pet owners now take more photos of their animals than of their family.

The book you hold in your hands contains three hundred images, spanning the whole development of the medium since its public arrival in 1839. Just a handful, then, from the millions made that century and beyond as photographic culture encircled the globe. The images brought together here, whether a glimpse of nature's wonder or a pet's everyday peculiarity, are fractions of a second in the span of photography. If a calculation were necessary, we could say this entire collection represents less than three minutes of life on our remarkable planet. From over 180 years of photographic image-making, we have here barely 180 seconds. Definitive is impossible.

So, we must be selective. Think of this book as a different kind of family album. An album of animals that is at times confusing, challenging, mundane and beautiful. We include important images and everyday images. This collection is, in part, a history of the technological innovations that

Daniel Szalai, *Novogen*, 2017
–
Szalai shot a series of portraits of Novogen white hens, a breed of chicken genetically engineered to lay eggs for the production of vaccines and other pharmaceutical products.

have made new kinds of image possible, as well as a reflection of changing attitudes towards animals. I'm concerned with the *why* of a photo, not just the how. It's a gathering that speaks of human ingenuity, technical expertise, daring and curiosity. It's a mirror of ongoing fascinations, yes, of broadening understanding and appreciation of animals, but of our continuing ignorance too. Of the *not seeing*, of denying realities, of the deliberate looking away. We are making more and more images of animals every year and yet animals are suffering on our planet as never before in human history.

* * *

We humans have caused unrelenting damage to the natural world. The UK State of Nature report in 2019 found one in seven UK species to be at risk of disappearance. Research shows that North America has lost three billion birds – around 30 per cent of total populations – in just fifty years. As extinction rates climb, biologist E. O. Wilson warns that were are entering the 'Eremocene', the Age of Loneliness. Mammals, birds, plants and insects are vanishing in staggering numbers. Humans make up just 0.01 per cent of all living beings, but have destroyed 83 per cent of wild mammals since the dawn of civilization. Or consider the distribution of mammals on our planet today: over 60 per cent are livestock, mostly cattle and pigs; 34 per cent are humans; and just 4 per cent are wild animals. When it comes to birds, 70 per cent are farmed poultry.

Globally there are over twenty-one billion chickens, feeding our insatiable demand for eggs and fried nuggets. These are not the wild forest

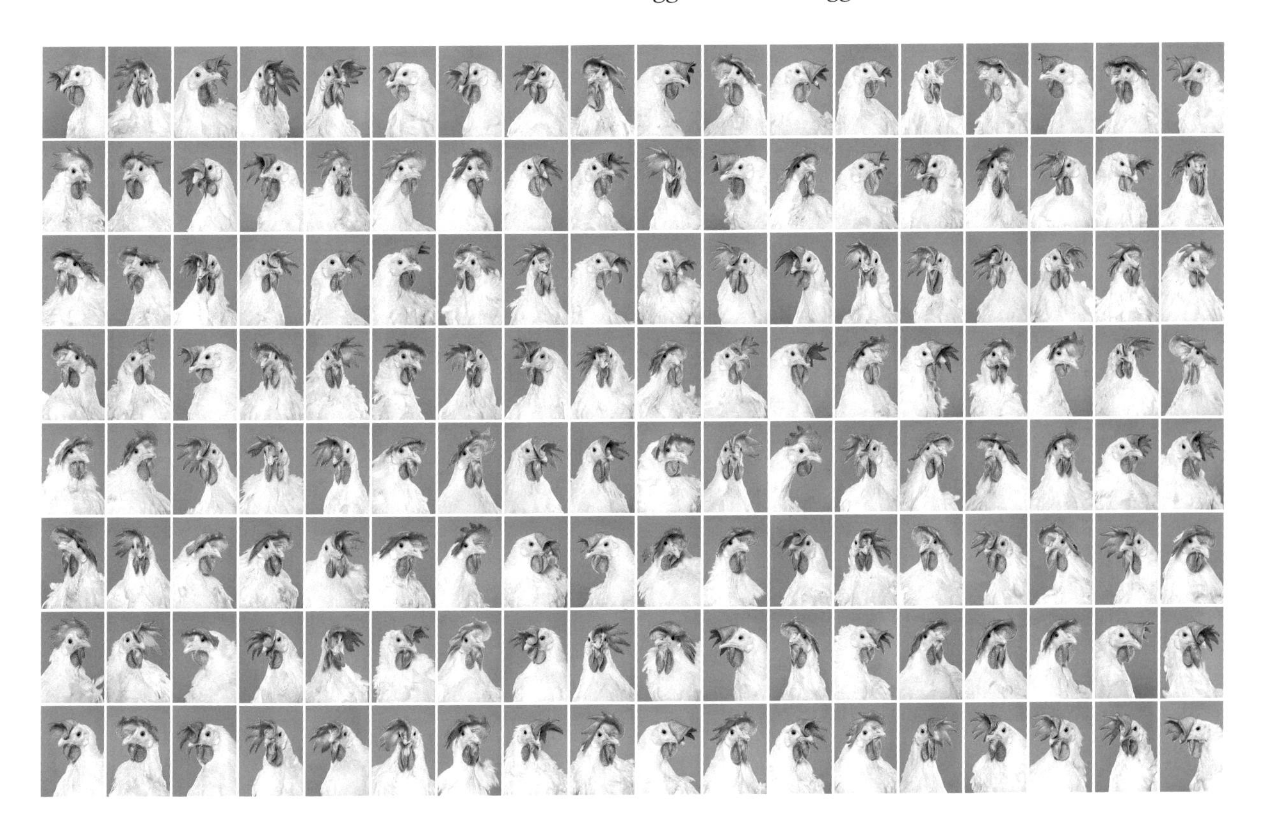

birds that were first domesticated by humans about 8,000 years ago, but new genetically modified, corporately controlled, patented inventions with shorter lives and bigger breasts, doomed to exist in factory farms never seeing sun, sky or blade of grass. Photographers, film-makers and activists are working night and day to try and draw attention to, and challenge, the insane logic of our food and pharmaceutical industries, and yet still these issues are unknown and of little interest to the majority of consumers. We should all be concerned. It's a colossal lack of consciousness and the numbers are overwhelming.

Yes 'pictures of infinite beauty' – recalling Ruskin, if we take time to look closely – but we must also be very much aware that our precious planet is not infinite. Scientists talk of 'biodiversity loss', which, when you think about it, is such an insufficient euphemism for the annihilation of non-human beings. Now is surely the time, long overdue, for us to stand up and take notice of these unpalatable truths. To use photography as a weapon. 'The truth is the best picture', Robert Capa famously remarked, 'the best propaganda'. The earth is the only home we have. This crisis demands a new kind of creativity.

Many photographers are motivated to take pictures of animals because of their profound distress at the state of the world's wildlife, but they can – and must – also make work to draw new audiences to the wonders of nature and to celebrate success stories, giving us hope in the darkness. As Richard Misrach has written, 'beauty can be a very powerful conveyor of difficult ideas. It engages people when they might otherwise look away.'

'What's happening is that people are making a billion photographs a year of their cats, frequently with the cats wearing costumes. Do you think I should be doing shows of cat photography?'

John Szarkowski

* * *

A baby dolphin is pulled from the sea as a crowd of sunburned tourists gather. They jostle for selfies at the water's edge, laughing, pushing, smiling for the cameras. It's an endangered breed, a franciscana, known as a 'ghost dolphin' by conservation experts who fear for its future. A photograph of the dolphin was published by *TIME* magazine in 2016 in a feature describing the year's most influential animals. After throwing the dolphin around for a while, the tourists in Argentina left it to die in the sand.

The roll call of animals in the *TIME* article included many that had been given names and have, to varying degrees, intriguing cultural histories – victims of our fascination and our ignorance. There was Cecil, an African lion hunted down by a Minnesota dentist; Tilikum, a captive orca, seen in the remarkable 2013 film *Blackfish*; Uncle Sam, an American bald eagle forced into a photoshoot with a president; Suraj, a Hindu temple elephant rescued from years of abuse; a Grumpy Cat with millions of fans; Feten, a bull who defeated the matadors in Madrid's Las Ventas ring but was still slaughtered by sword; Otto, a skateboarding bulldog famous in Peru; and

Roxana Dulama, Pitzush the fashionista cat, 2016
–
Pitzush, a Nebelung breed, was adopted by Dulama, a Romanian dentist and blogger. Adding headgear and using Photoshop to create fake eyelashes, she posts her pictures to Instagram for Pitzush's followers to enjoy.

Eastman Kodak Company advertisement, 1900
–
George Eastman's Kodak camera hit the market in 1888, with its newly invented photographic film. At last photographers were freed from the burden of heavy boxes of plates and chemicals. Portable and easy to load, it proved an instant hit.

a group of goats genetically modified in China with bigger muscles and longer hair to increase meat and wool production. And so on.

Equally saddening, consider these headlines: 'Man trampled to death by elephant', 'Man mauled by a bear', 'Woman attacked by a jaguar in Arizona', 'Man drowned by a walrus in China'. And what do these tragedies have in common? The people were all trying to take selfies. Nature photography is a dangerous business sometimes, but this is crazy.

Is it ignorance that compels people to risk their lives like this? Is it because nature is somehow less real to people these days? Or, as some commentators have suggested, is it love? They *love* the animals so much. Perhaps, more to the point, people are starting to love photographs too much. Whatever the motivations, it's a troubling development in our culture. And it is behaviour that is ultimately bad for animals, whether in the wild or already in captivity. The animals in question are usually punished in some way, or further harassed, or caught and relocated or even hunted down and killed. All because of the human desire to take a photo.

* * *

Why We Photograph Animals encourages practitioners to think more about the images they make. Of course, the impulse and the intentions behind any creative act are varied. Many early photographers were hunters first, before swapping their guns for cameras as they became aware of the need to protect the wildlife they loved. Motivations evolve during a career, that is natural, and images can be made to work in different ways too.

Through photography we can explore the social, symbolic, scientific and enduring aesthetic intrigue of animals: from portraits of beloved pets, to studies of wild animals in their natural habitat, or even documents of human cruelty against them; from curiosity to wonder, love, anger or hope. Evoking beauty and joy, fear or fury, or even inspiring action in audiences around the globe, many of these images of animals have changed the way we think and feel about the environment. And some are just plain cute, and there is no harm in that.

New techniques and technologies are helping to create images that only a few years ago would have seemed impossible: in high-definition digital, with camera trap and drone and all manner of innovative, remotely operated gear. But we also have images shot on smartphones that are every bit as masterful as those captured on glass plates. The digital age has changed the photography of animals more than any other genre. Particularly so underwater, where you now don't need to surface to change film after just thirty-six frames. The speed of digital capture has resulted in shots

of animal behaviour never before seen, while digital sensors have allowed photography in the lowest light conditions.

Neglected in most photographic surveys, animal photography deserves its place in a new art history. Top photographers are using their talents to bring attention to the challenges our environment faces and their images are engaging in some of the most pressing issues of our time.

Photography enables people to experience something of the wonder and life of animals. 'No one will protect what they don't care about', as David Attenborough has told us many times, 'and no one will care about what they have never experienced.' At its best, photography can broaden our minds and awaken our senses: helping to uncover the intricate complexities of animals and the millions of ways in which they live. But what was it that Ruskin said again? If only we have the eyes to see.

(above) John Downer Productions, animatronic 'spy creatures', 2017

–

Many years in the making, John Downer's *Spy in the Wild* TV series used a host of remote-controlled animatronic creatures with ultra-high-definition cameras for eyes, to give viewers up-close and personal footage never seen before.

(opposite) Cherry and Richard Kearton photographing bird nests in high hedges, Yorkshire Dales, 1896

–

As well as using the latest equipment of the time, the Keartons ingeniously adapted new kit, like extendable tripods and canvas hides.

(overleaf) Adam Oswell, Khao Kiew Zoo, Thailand, 2019

–

Tourists watch an Asian elephant forced to perform and swim underwater for their photographs. Oswell's award-winning image encourages us to ask questions of this exploitative industry.

HISTORIES

Infinite Storm

‘When we contemplate the whole globe as one great dewdrop, striped and dotted with continents and islands, flying through space with other stars all singing and shining together as one, the whole universe appears as an infinite storm of beauty.’

John Muir

THOUSANDS OF PEOPLE come to see him. He had been pulled from the banks of the Nile as a baby, caught, bound and caged. His parents shot. But that isn’t the story they are told. It is enough that he has arrived: a gift from the Viceroy of Egypt, a present fit for Queen Victoria. A paddle steamer brought him to England and he is given the name Obaysch, after the island in the river where he was captured.

He is the first seen in Britain since Roman times, it is said, and now here he is, held as an exotic prize, on show in the centre of an increasingly global city. Some write of ‘hippomania’ spreading across London. ‘Everybody is still running towards Regent’s Park’, describes *Punch* magazine in 1850, and ‘he repays public curiosity with a yawn of indifference.’ One of his many admirers is Don Juan the Count of Montizón, a Spanish aristocrat and keen photographer. We may never really know *why* he photographs the hippo that day. We assume he enjoys watching animals, but maybe he feels sorrow for their captivity. Perhaps making images is simply a hobby? The hippo’s lazy repose is probably another factor – at least it doesn’t move much and blur the shot, which also appears to have been taken inside the cage. The subject is as much the crowd as it is the beautiful beast before them; it seems the people are the ones caged in.

The industrializing nations are developing new technologies and a taste for the exotic. Advances in transportation are shrinking the world, and those with the freedom and wealth to travel are having experiences that would have been unimaginable to a previous generation. Yet for most, life in the mid-century city is tough, dirty, difficult, dehumanizing. Identities are lost in a swathe of progress: a tumult, an upheaval, a disorientation. And, for Obaysch the hippo? Well, he is fed twice daily, as the noisy crowds pour past his enclosure. Songs are sung about him, music composed and products sold, and he is photographed again and again as the years pass. Perhaps Obaysch and his hippo family are some of the lucky ones? Not shot, stuffed and mounted, like millions of other animals that century. Lucky, some say, to be living in a zoo. Perhaps. But what kind of life is that?

Don Juan, Count of Montizón, ‘The Hippopotamus at the Zoological Gardens, Regent’s Park’, 1852
–
Obaysch was London Zoo’s first hippo and was said to have drawn up to 10,000 visitors every day.

Photography's first big year is 1839. Simultaneous announcements of two processes for fixing an image directly from nature, using optical and chemical means, cause a wave of excitement in England and France, and the news spreads throughout the Western world. After many years of experimentation, Louis-Jacques-Mandé Daguerre, a Parisian stage designer, declares the perfection of a process that produces an image on a silver-coated copper plate. Across the Channel, an English gentleman, William Henry Fox Talbot, reveals a positive-negative technique on paper that is destined to become the basis of modern photography. Both are united by a profound conceptual advance, ideally suited to the changing time: the face of nature can be recorded accurately, measured, possessed, captured for posterity.

'No human hand has hitherto traced such lines as these drawings display; and what man may hereafter do, now that Dame Nature has become his drawing mistress, it is impossible to predict.'

Michael Faraday

Despite its initial popularity, the daguerreotype suffers from a series of basic disadvantages that mean it's quickly overtaken by other methods of photographic recording. It's often described as photography's false start. Exposure times were just too long and though this is eventually reduced to minutes rather than hours, it still places limitations on the choice of subject. Animals are pretty much off limits in the earliest years, they simply moved around too much, though some rare examples can be found.

Talbot's positive-negative process is the true photographic game changer. It allows multiple copies to be made from a single image, ensuring the medium has the reach and prevalence to make it a genuine mass cultural product. Talbot had begun experimenting while on his honeymoon in Italy in 1833 but had kept his research secret. Daguerre's announcement

Louis Daguerre, *Shells and Fossils*, 1839

–

In the early days of photography, it was helpful to have subjects that would remain motionless during long exposure times. The ammonites in the centre, which we now know as long-extinct cephalopod molluscs, were believed by early collectors to be coiled-up snakes that had been turned to stone. Many of these specimens are millions of years old, an echo of the world long before the imprint of man.

Joseph-Philibert Girault de Prangey, 'Near Alexandria, in the Desert', panoramic daguerreotype, 1842
–
Travelling extensively through Greece, Syria, Palestine and Egypt, de Prangey visited ancient ruins and photographed many of these wonders for the very first time.

drew him to reveal his own: the 'Talbotype' or *calotype*, meaning 'beautiful picture'. He uses sensitized paper as his base and produces, in the camera, a negative image that is later placed in a printing frame with a new sheet of sensitized paper behind it. After exposure, a print is produced by contact and then fixed in common salt. Talbot shows his images publicly for the very first time at the Royal Institution in London on 25 January 1839. His friend the eminent man of science Michael Faraday announces the parallel discoveries. Faraday observes that 'no human hand has hitherto traced such lines as these drawings display; and what man may hereafter do, now that Dame Nature has become his drawing mistress, it is impossible to predict.'

An early use for the infant art of photography is to make a permanent record of prepared specimens. Using a solar microscope, a powerful beam of light could be concentrated on the slide. Talbot is able to image the wings of a lantern fly and in 1844 Léon Foucault uses this process to capture images of human milk, mouse sperm and the blood cells of a frog. Hippolyte Fizeau is also experimenting with the new medium, creating 'macroscopic views' of bed bugs.

As 1839 progresses, despite miserable weather, Talbot is able to expand his portfolio. He prepares for his second and what would be his largest exhibition: the annual meeting of the British Association for the Advancement of Science, which brings together up to 1,500 amateur scientists from throughout the country. Talbot gathers a substantial exhibit – ninety-three items in all – to demonstrate the versatility of his work on paper. In time, he also creates a book, *The Pencil of Nature,* which appears in six parts beginning in June 1844. Not only is it the first book on sale to contain photographic images, it also provides, in a sense, a visual language for a new generation of practitioners. It is both an advertisement for the art and its manifesto – a call to arms.

A few intrepid photographers begin to make work overseas. In 1842 Joseph-Philibert Girault de Prangey embarks on a journey through the Eastern Mediterranean, hoping to become an expert in Islamic architecture. After three years he returns home to France with more than one thousand

Daguerreotypes, 1850–56

–

By the early 1850s, most major cities in Europe and America had studios specializing in daguerreotype photography. Holding still for the required exposure time was not easy for humans, let alone a dog or a mighty eagle, whose movement accounts for the blurring. From family pets to prized livestock, to an elephant in a travelling circus, these are rare treasures indeed.

ST. LOUIS

daguerreotypes. His efforts could be considered the world's first photographic archive. In the desert near Alexandria he captures an image of a laden camel, possibly one that carried him that day. Though blurry, its form is clear: one of the earliest living animals rendered in a photograph.

Later, in 1852, Ernest Benecke, the son of a German banking family, also heads to the Mediterranean. Though few of his images survive, there is an intriguing shot of a crocodile on the deck of his little local boat in Upper Egypt. It has been hauled aboard and its belly opened up for inspection. Considering that it's more than forty years before hand-held cameras and fast films make the snapshot genre possible, this is a fascinating, candid shot. Though faces are blurred, and the subject grisly, we are instantly transported.

John Franklin's expedition in 1845 is the first to take camera equipment up into the Arctic but both his ships disappear off the edge of the map, and no men live to tell the tale. When Edward Inglefield sails for the north in the summer of 1854, *The Times* reports that he has with him 'a most complete series of the articles used by photographists for depicting nature as seen in the polar regions', but Arctic animals are elusive. A small selection of his photos on glass survive, created as his ships move up the west coast of Greenland. A dead seal appears in a portrait of a hunter with his skin kayak; the seal is inflated as a float, tethered to the end of the harpoon line. But still there are no living animals in these images.

* * *

The year of Daguerre's death, 1851, witnesses the beginning of a new period in photography. The great invention, which supplants all existing methods, is Frederick Scott Archer's 'wet collodion' process, so called because the

Samuel Alexander Walker, *carte-de-visite* of actress Nelly Moore, 1862

–

Highly collectible, posted with letters and exchanged among friends these 'calling cards' were, in a sense, an early form of social media.

Ernest Benecke, 'Autopsy of the First Crocodile Onboard', in the upper Nile, Egypt, 1852

–

The more famous photographers who travelled to Egypt shortly before and after Benecke focused almost exclusively on the ancient monuments and landscape. Benecke, instead, documented the contemporary world with a keen eye, presenting intimate and unaffected portraits of the region's inhabitants.

photographer has to coat the plate, expose it and then develop it all while it is still wet. It is the fastest photographic process yet devised and is immediately popular, despite being a tricky proposition. Taking a wet-plate photo still means carrying a darkroom and a case of noxious chemicals, making remote location photography difficult. Up to about 1880, the collodion process is still the most widely used technique. Even today, where digital has not totally won out, processors continue to use wet collodion.

Albumen (egg white) is found to be good for coating positive paper and, once gold chlorine toner is devised, photographs appear in rich detail. Soon albumenized positive paper can be bought already prepared, and the portrait business booms into a mass market. Such is the demand for albumen in the 1850s that factory farming of hens is introduced, and when the desire for cheap *carte-de-visite* pictures is at its peak, over half a million eggs are used annually by one London firm alone. Thirty years later, after Kodak revolutionizes amateur snap-shooting, the US photographic market is devouring 300 million eggs a year.

The relaxation of Talbot's patent in July 1852 and the gradual perfection of the collodion process at last make photography a popular pursuit. In 1854 the London School of Photography opens its doors, providing teaching and instruction. By 1861 over two dozen photo societies have been established, with enthusiasts all over the country. Portraits drive the industry, but the uses of photography have expanded considerably. In archaeology and architecture and the new sciences of natural history, meteorology and microscopy, photography becomes an essential tool for recording. Even Queen Victoria has a darkroom constructed at Windsor Castle and is said to have become very skilled in the 'black art' under the guidance of Dr Becker, Prince Albert's librarian.

William Bambridge is hired as official photographer to the royal household. Beyond what must have seemed like endless family portraits, he photographs the queen's dogs and deer in Windsor Park. In 1860, her miniature portrait secures the *carte-de-visite*'s fashionable status and encourages a frenzy of production across London. The queen becomes an avid collector of photographs and, in time, prints and engravings of her many photographic portraits are displayed in homes, school halls and railway stations throughout the British Empire.

John Dillwyn Llewelyn, *The Lewitha*, albumen silver print, 1856

–

A cousin of Talbot, Llewelyn experimented making images in the woods of his estate in Penllergare, Wales using various stuffed specimens from his collections: a heron, an otter and this red deer stag.

In 1858 the *Stereoscopic Magazine* is founded, and each issue offers subscribers sets of double images. With images of the Canadian forests, the Alps and the Nile, spectacles of nature are commodified: all these and more can be enjoyed from the comfort of one's armchair. It's a liberating invention, giving people a chance to escape the everyday, to dream and travel the world vicariously, and desires are fed by an abundance of imagery. The 1860 catalogue of the London Stereoscopic and Photographic Company lists over 100,000 cards in stock, with more than half a million views already sold. That same year, newspaper readers learn of an American invention that claims to print over 12,000 stereographs an hour. This is a brave and brash new visual world.

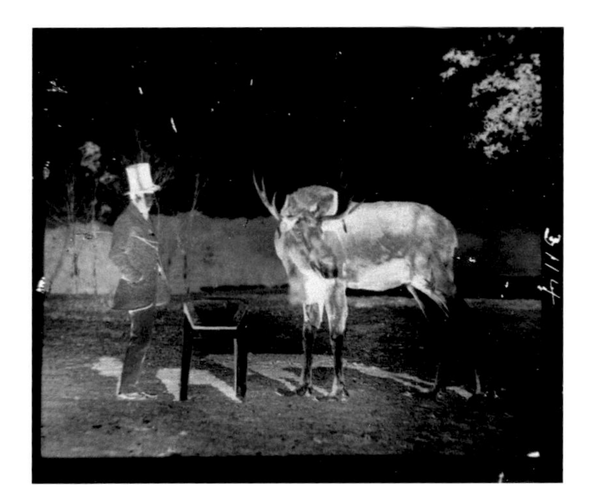

(above) William Bambridge, 'Elk in Windsor Park', glass plate negative, 1854
–
Bambridge was appointed photographer to Queen Victoria and for more than a decade created images of her children, pets, royal hunts and visitors to the castle.

* * *

In America, photographers are discovering landscapes wilder than any seen in Europe. In the summer of 1861, Carleton Watkins sets out to photograph Yosemite. Sketches and awed descriptions of Yosemite's grand views had reached the east in the mid-century, but nothing provokes public reaction like Watkins's shots, which are soon exhibited at a gallery in New York. 'The views of lofty mountains, of gigantic trees, of falls of water ... are indescribably unique and beautiful,' the *New York Times* reports. In 1868, he is awarded a medal for landscape photography at the Paris International Exposition and in 1876 he shows his work at the Centennial Exposition in Philadelphia. In his lavish 'Yosemite Art Gallery' he displays large 'Pacific Coast' views with a 'thousand of the best stereographs' to thrill and inspire visitors.

Animals become one of the key subjects driving the technical development of photography. Prior to the 1890s, technological limitations mean that animal photography beyond pets and livestock mostly took place with captive or even dead animals, as we have seen. By 1865 Frank Haes is having success creating shots of animals in London Zoo: he finds the sleeping hippo easiest to capture, but soon moves on to a wolf and an eagle. He succeeds with a zebra too and a seal, despite punters throwing fish to

(below left) Carleton Watkins, sea lions at the Farallon Islands off San Francisco, c. 1870
–
Watkins travelled widely around California in pursuit of interesting and profitable subjects. Successful for a while, his studio was destroyed in a fire and he later died penniless in an insane asylum. Just two months later, President Woodrow Wilson established the National Park Service, a steward for the sublime places Watkins had shown to the nation.

distract it as he was trying to get his shots; he goes into an inner den with the lion keeper; he almost gets his leg chewed by a Tasmanian tiger. But in the quest to photograph wild nature, photographers push cameras to new levels. In 1867 Timothy O'Sullivan is dodging 'voracious and poisonous mosquitoes' in Nevada, and using a horse-drawn wagon and boats to access the most remote parts of the US. John Hillers is in the Grand Canyon. Laton Huffman records the last herds of bison still roaming the open range in the 1880s. Later, dry gelatin plates, high-speed shutters and telephoto lenses all increase what is possible.

Take Reginald Lodge, for example, who is sometimes credited with having taken the first photograph of a wild bird, a northern lapwing on its nest, in 1895 using a tripwire placed over its eggs. Lodge and his young assistant, Oliver Pike, used to haul a 12 × 10-inch plate camera around the English countryside in a wheelbarrow. Lodge was a founding member of the Zoological Photographic Club in 1899 and he was awarded the first ever medal for nature photography by the Royal Photographic Society. Pike, meanwhile, becomes so frustrated with the cumbersome wheelbarrow approach that he develops his own 'Birdland' camera. It is portable enough to be used for stalking and small enough to fit inside another new innovation: static photo-hides or 'blinds', as a means of getting closer to wild animals.

Another notable early member of the Club is Richard Kearton who, along with brother Cherry, are pioneers of both still and cine photography. The brothers create some of the earliest surviving nature photos, such as the nest and eggs of a song thrush, which they expose in April 1892. The first book about animals to be illustrated throughout with photographs is their *British Birds' Nests*, published the same year that Lodge takes his celebrated lapwing photo. In 1897 they reveal the secrets behind their innovative techniques and equipment in *With Nature and a Camera*. Their efforts send hosts of amateur photographers out into the field to study wildlife for themselves.

* * *

Expeditions to the polar regions often took photographic equipment, and image-making was a means of

(above) Laton Alton Huffman, 'After the Chase', North Montana, 1882

–

Unlike the Native Americans, European Americans killed bison in large numbers and only took the hides and trophy-heads, usually leaving the rest of the animal to rot. This image was later used to illustrate the plight of the American bison and the need for wildlife conservation.

(right) Reginald Lodge, lapwing incubating its eggs, 1895

–

Long thought to be one of the first photographs of a wild bird, we now know of others, such as Caleb Newbold's rockhopper penguins in 1873, or Ottomar Anschütz's action studies of storks in 1884.

John Dunmore and George Critcherson, 'Hunting by Steam in Melville Bay', 1869
–
Photos from this voyage were later published in William Bradford's *The Arctic Regions*. 'The party after a day's sport', the text details, 'killing six Polar Bears within twenty-four hours.'

recording and seeing landscapes and peoples in new ways. It was a technology of power and control, as much as artistic curiosity. Early images of wild animals can be discovered among the documentation created and disseminated by these expeditions. In 1869 American artist William Bradford charters a steamer named *Panther* to cruise to Labrador and Greenland. He has been to the Arctic before, but this time hires two Boston studio photographers, John Dunmore and George Critcherson, to join his crew. Bradford sees the value of photos as a visual aid when creating his epic paintings, as gifts for potential patrons, as well as for publicity material to drum up sales. His photographers work hard under the midnight sun. Proceeding northward into the ice, they're rewarded with their most exciting opportunity yet. A mother polar bear and her two cubs are spotted on the ice:

> At this moment the photographers came rushing on the deck demanding the right of a 'first shot'. Quick as a flash the camera was down and focused, a slide with a little hole in it was dropped before the lens, and the family group of bears was taken at a distance of about two hundred yards. To accomplish this feat required the very first degree of enterprise and skill. The camera was stationed upon the top-gallant forecastle, and the impression was obtained while both ship and bears were in motion.

There is nothing romantic about what happens next. The bears are dispatched by rifle and both hunters and prey are photographed in front of the ship. Sad as the story is, the images are important: the earliest photographs of truly wild animals that survive. It's just a shame the polar bears didn't.

Despite these images, it's a picture of a small group of penguins that is often cited as the first wild animal photograph. It comes from the voyage of the *Challenger* in 1873, a remarkable scientific cruise organized by the Royal Society under the command of George Nares and naturalist Charles Thomson. It lays the foundations for modern oceanography, with pioneering dredging of the ocean floor to map its geology and the collection of thousands of biological specimens. It is fitted out with an onboard laboratory for the naturalists and a small darkroom too. Corporal Caleb Newbold is tasked with photography, and he has been trained by William Abney at the School of Military Engineering in Chatham. There is sadly no direct documentation of the equipment used, but it can be assumed that it follows Abney's recommendations, likely the wet-plate collodion process.

On 16 October 1873 Newbold photographs rockhopper penguins on Inaccessible Island near Tristan da Cunha in the South Atlantic, certainly the first time penguins are photographed in the wild. There is little time to celebrate this milestone though: Newbold jumps ship in Cape Town, leaving the photo mission in disarray. He is frantically replaced by Frederick Hodgeson, who suffers terribly from seasickness, and then Jesse Lay, who stays on all the way back to England. Among many wonders seen for the first time, Hodgeson photographs nesting albatross at Marion Island in the Indian Ocean and a termite nest at Cape York. Photography serves a number of other roles on the voyage too: for recording charts, preserving sketches and artworks, and photo prints are even offered as gifts when visiting dignitaries ashore.

Caleb Newbold, 'Northern rockhopper penguins amongst the tussock and boulders', Inaccessible Island, 1873

–

Many animals were photographed for the very first time during the voyage of HMS *Challenger*.

George Nares leaves the *Challenger* voyage halfway through, taking command of another naval expedition tasked with exploring the Arctic. In the ships *Discovery* and *Alert*, he steams through the ice to Ellesmere Island between 1875 and 1876, attaining a point further north than any other to date. By now the development of dry-plate photography enables pictures to be taken in intense cold. The pre-coated plates can be exposed then developed back on board. George White and Thomas Mitchell secure a series of stunning images that are crucial to the expedition's reputational success. Dispatched home on a supply ship, the photos are used for engravings in lavish supplements in *The Graphic* and the *Illustrated London News*. Despite the huge distance involved, photography has become a tool of reporting and Arctic images feature on a weekly basis.

George White and Thomas Mitchell, 'Franklin Pierce Bay, 79°25'N', 1875

–

The ice quartermaster of HMS *Alert* poses with the walrus he had just harpooned.

Despite failing to reach the pole, and debilitated by scurvy, the explorers are welcomed home by an adoring crowd. Images circulate widely in the newspapers, sometimes even appearing on the front page, they adorn souvenir pamphlets and they inspire specially painted sets of magic lantern slides. There is a glimpse of animal life – though lifeless, a harpooned walrus and the head of a musk ox – photographed as hunting trophies. Ships, ice, sledging and scenery are the mainstay. 'Photomechanical reproductions' are sold as prints, while a full box set of photos, bound in red silk, is given to Queen Victoria to add to her growing collection.

By 1884 the first flexible negative film is produced, necessitating another dramatic change in equipment. A few years later, George Eastman presents the first Kodak camera, a name he chooses because it 'can be pronounced anywhere in the world'. It is small, with a single shutter speed of 1/25s and a fixed focus, but it is revolutionary in two important ways: being both cheap and easy to use. It is a snapshot camera for the masses, with the now famous slogan: 'you press the button, we do the rest'.

With the arrival of the Kodak Brownie at the turn of the century, and its sensational price of just one dollar, photography is democratized. It becomes the most mobile and accessible of all visual forms. During that first year alone more than 100,000 are sold, half of them in Europe. 'Now every nipper has a Brownie', the photographer Alvin Coburn despairs, 'a photograph is as common as a box of matches.' Of course, not all photographers want to hand the finishing work over to a factory,

(right) Eastman Kodak Company, Advertisement, 1893
–
'Peary's sledging trip with a Kodak across the Greenland ice cap', the copy runs, secured 'over 2,000 superior negatives in a region hitherto unexplored.'

but what the mass-produced Kodak does achieve is to give the photographer a choice. The enthusiastic amateur photographer, and professional cynic, George Bernard Shaw, observes 'the photographer is like a cod which lays a million eggs in order that one may be hatched'. From this point on, cameras can be carried anywhere. For a hunting trip to the lakes, a holiday to the seaside or a long voyage to exotic lands, the Kodaks are ready to capture it all.

The explorer Robert Peary takes advantage of this lightweight camera gear and, pleasing his sponsors, manages many photographs of the Arctic and its peoples. Based on his experiences he even writes a do-it-yourself manual, *The Kodak at the North Pole*, certainly the first guide to polar photography. Eastman prints it as a publicity piece. There are other explorers and other cameras, of course. One of the best is Herbert Ponting, the official photographer on Captain Scott's now notorious final expedition, which sets out for Antarctica in 1910. Ponting describes himself as a 'camera artist' and takes with him some of the first colour plates produced by the Lumière factory at Lyons.

(below) Herbert Ponting, publicity souvenir, 1912
–
In this montage of Adélie penguins, Ponting poses with his Prestwich Model 5 Kinema Camera. At the very bottom we see his mobile photographic set up: a little wheeled trolley pulled by two dogs. 'The Kinematograph, properly applied', he later wrote, 'is the greatest educational contrivance ever conceived'.

The expedition is certainly the best equipped, from a photographic standpoint, of any that has yet left England. The cameras for the sledging parties are specially made 'Sibyls' by Newman & Guardia, with attention paid to strength and compactness. The 'cinematograph' camera is built by Newman & Sinclair with unique Tessar lenses and pioneering reflex focusing to ensure smooth running. Though Ponting is seasick most of the way south, at the sight of icebergs, he claims that he is the happiest he has ever been in his life.

Conditions at base camp are tough. Water freezes in Ponting's tanks, flesh sticks to frozen metal and outdoor photography means a constant risk of frostbite. But he sticks to his task and creates evocative studies of icebergs, portraits of dogs, horses and men, and superlative images of expedition life at the edge of possibility. Scott praises him many times in his journal: 'His results are wonderfully good, and if he is able to carry out the whole of his programme we shall have a photographic record which will be absolutely new in expeditionary work'. Scott is right to trust in his skill: Ponting exposes around 25,000 feet of film and 2,000 photographic negatives during the expedition, some of the most striking images ever made in the Antarctic. In their beauty and composition, they have rarely been equalled.

* * *

I sometimes ask my students: what do storm waves at Dover, a cockfight, the final stages of the Epsom Derby and a boxing kangaroo have in common? Answer: they are the very first 'nature' subjects captured in moving images, or what comes to be understood as modern motion pictures. The year is 1895: at first, films are silent and black and white; later sound and colour are introduced. And of course, it is the moving image that ultimately surges ahead in this visual culture. As in photography, the major difficulties are technical. To haul a motion-picture camera, developing equipment and a sufficient supply of film is a huge challenge, but photographers innovate and their sponsors rise to the challenge: there are fortunes to be made.

'The small attempts at architecture have swelled into monumental representations of a magnitude, truth, and beauty which no art can surpass – animals, flowers, pictures, engravings, all come within the grasp of the photographer.'

Elizabeth Eastlake

Eadweard Muybridge's experiments in the 1870s begin to extend the photographic gaze: he creates a sequence of a galloping horse to prove all its hooves leave the ground at the same time, helping its owner win a bet. His later studies of animal locomotion are a sensation. In the years that follow he applies his pre-cinematic techniques to more horses, dogs, pigs, pigeons, even wild deer. These deer in fact could be the first true wildlife to be captured in any kind of motion-picture process.

Muybridge takes his motion-photo technique to the Philadelphia Zoological Gardens in 1884 and 'films' a tiger as it attacks and kills a buffalo. The confrontation is staged for the camera: the first, but certainly not

(below, left) Eadweard Muybridge, *The Horse in Motion*, souvenir card, 1878

–

In 1872 Muybridge was commissioned to create a photo sequence of a galloping horse to prove all its hooves left the ground at the same time.

(opposite, above) Osa Johnson riding a Zebra near Mount Kenya, 1921

–

Osa rides a domesticated zebra named Bromar in a shot widely used for publicity for the film *Trailing African Wild Animals* (1923). A huge number of animals were killed making the 'real nature' film, yet the American Museum of Natural History gave them funding to return and produce more.

(opposite, below) William Longley and Charles Martin, a hogfish in the Florida Keys, 1926

–

To achieve this image, the camera was encased in a waterproof housing and a raft laden with highly explosive magnesium flash powder floated above to provide the burst of light needed.

the last, episode of fakery in wildlife film-making. Muybridge continues his work at the University of Pennsylvania and by 1887 publishes *Animal Locomotion*. In 1893 he gives 'zoopraxograph' demonstrations at the Chicago World's Fair, where spectators see 'flocks of birds fly across the sky with every movement of their wings perceptible.'

Ottomar Anschütz is also experimenting with action photography. His focal-plane shutter plus sensitive photographic negative material is the key. A series of flying storks from 1884 are among the earliest action shots, and his studies of horses are used by artists. He develops machines to show a sequence of photos on glass plates and, later, images printed on celluloid or mounted on card. His *Elektrischen Schnellseher* (literally, the 'Electrical Quick-Viewer') is a crowd-pleaser at the Chicago World's Fair in 1893. Sequences include dogs and horses, a flying stork, a leaping goat, a camel running and a man riding an elephant.

In 1903 Charles Urban starts his 'bioscope expeditions' to capture footage in remote parts of the world, featuring dramatic scenery, indigenous customs and exotic wildlife. He also funds a series of 'micro-bioscopic' films to reveal the dramatic unseen world of nature, such as Martin Duncan's *Cheese Mites* and *Circulation of Blood in a Frog*. In 1907 Dr Adam David takes a cameraman on his hunting safari along the Dinder River and produces perhaps some of the earliest moving pictures of wildlife in Africa. By the 1930s, when Martin and Osa Johnson are back in Africa to make an 'aerial safari' from Cape Town to Cairo shooting their blockbuster movie *Baboona*, they are able to cover thousands of miles. They use a pair of Sikorsky boat-planes emblazoned in zebra and giraffe print, meeting the needs of logistics and publicity in a perfectly on-trend package.

In 1913, John Ernest Williamson's illuminated photographs of the depths of Chesapeake Bay inspire him to try motion pictures too. He makes the first ever underwater film in clear waters off the Bahamas using an underwater observation chamber, which he calls 'the photosphere'. Underwater colour photography is born later, in 1926, with a shot of a hogfish, photographed in the Florida Keys. The pioneers William Longley and *National Geographic* staff photographer Charles Martin equip themselves with waterproof cameras and highly explosive magnesium flash powder for underwater illumination. But it is not until much later, in the 1950s, that underwater photography really comes of age. In 1956, photographer Luis Marden accompanies the now legendary ocean explorer Jacques Cousteau on a voyage from Toulon to the Suez Canal aboard Cousteau's ship, *Calypso*. By journey's end, Marden has 1,200 photographs, the largest collection of underwater colour images yet achieved.

* * *

If we return, finally, to things polar: Frank Hurley's pioneering 35 mm footage, shot during Ernest Shackleton's attempt to cross Antarctica between 1914 and 1916, is some of the best of its kind. When their ship *Endurance* sinks, Hurley is able to save some of his film and glass-plate negatives and an album of photographs he had already printed. He also manages to continue taking photos with his hand-held Vest Pocket Kodak after having to abandon his equipment at their ice-floe camp. Their voyage of discovery is turning into an epic struggle for survival.

But the best animal work, to my mind, comes from Hurley's first polar voyage, when he went south with Douglas Mawson in 1911. First stop is the remote Macquarie Island in the wild sub-Antarctic. Hurley is enraptured by the colonies of seabirds he sees there: 'had I sufficient plates and films', he writes, 'I could live here for the rest of my life'. Further south still, he pushes the limits of his kit to photograph the continent and its wildlife in impossibly difficult conditions, detailed at length in Mawson's book *Home of the Blizzard*. Of Hurley, Mawson later says: 'He is one of those to whom danger adds but a zest, one of those willing to undergo great hardships to accomplish an end'.

A photographer like Hurley has a range of qualities that continue to be relevant in the field in the present: tenacity and technical skill, stubbornness and flexibility, courage matched with creativity. As film speeds and equipment improve through the century, photographers are able to overcome the difficulties of motion, and go further and faster than ever before. And yet, going slowly, and quietly, remains vitally important. Some hunters swap their guns for cameras, while natural historians use new skills to reveal more of nature's myriad secrets. Artists and explorers, hobbyists and professionals, work the photographic image in different ways. Animal photographs embrace a wide variety of meanings and serve a multiplicity of functions: in experimentation and education, as proof and as record, in advertisement and spectacle. And, chief among all of these things, as an ongoing source of wonder and delight. An 'infinite storm of wonders', just as animals themselves were beginning to retreat from our lives.

(opposite) Frank Hurley, Adélie penguins after a blizzard at Cape Denison, 1912

–

Hurley created a trove of images on the Australasian Antarctic Expedition 1911–14, made famous in Douglas Mawson's *Home of the Blizzard*, and he returned five times to the continent.

(overleaf) Frederick William Bond, American alligators in London Zoo, 1928

–

Opening in 1926, the new Reptile House at London Zoo was designed by curator Joan Procter with the architect Sir Edward Dawber. With special glass, heating, dioramic backgrounds and focused lighting, at the time it was hailed as one of the most sophisticated animal enclosures in the world.

PROFILE

Florian Ledoux

FLORIAN LEDOUX *is a nature photographer and cinematographer from France. He was named SIPA Drone Photographer of the Year in 2018 and overall winner at Nature TTL Photographer of the Year in 2020. His work has been published in magazines including* Le Figaro, Paris Match, National Geographic *and* TIME. *He has also taken part in two major television projects, Disney Nature's film* Polar Bear *and the BBC series* Frozen Planet 2.

Can you remember the first photo of an animal that you took? A bee on a flower in my parents' garden taken with my first camera. At that time, I had no idea that nature and wildlife would become such a large part of my life.

Do you have a favourite animal photo that you have taken? My favourite image of my own is the one of the polar bear crossing the melting ice from above. Overall I love aesthetic and abstract aerial images that show wildlife in the wider environment and convey a meaningful message.

Have you ever risked your camera for a photo? I'm sure I have, but do we see it as a risk when we are driven by a powerful moment to create images? Being out there in the Arctic in winter is a challenge for humans. If you consider how far north it is – isolated – the risk is a part of the daily routine. On a snowmobile on the ice: you can fall through, ice can crack and drift and sailboats can get stuck or run aground. It is a constantly changing environment and the violence of the storms we face sometimes can be terrifying, but we push through, with control and knowledge, to get results. Creating great images requires going beyond your own boundaries, both mentally and physically.

Can you describe an animal photo that you would most like to capture? There are plenty of ideas for things I want to capture, but what I have learned over the years is to let nature guide you and keep your heart open. My way of creating images has always depended on what nature offers. If you make too many plans and try to script what you want to photograph you will most likely end up frustrated. With experience, you can learn how to read nature and a specific ecosystem to better anticipate a situation: observe, understand, adapt and then create. I learned that going with the flow of nature and the moment is more important for me in order to reach the magic zone. The zone where I feel closer to the environment around me, connected to nature. On every expedition, I discover a little bit more about nature and animal behaviours. This winter on the ice, I have witnessed some behaviours I'd never thought I would see – because nature and ecosystems are changing.

Has an animal photo forced you to examine your life? A photo of an orangutan in Borneo forced me to think a lot about my life and the impact we all have on the planet. The image shows the animal climbing a single tree left in the chaotic aftermath of deforestation. The message was a pretty clear one. It pushed me to study more about my footprint on the planet and how I can reduce it, but also to care more for the food I choose and to ask questions of the products I use.

It is often said that 'photography can change the world'. Do you agree? Absolutely, images can capture us and connect us emotionally. One image has the power to stand for those who cannot speak their own words. It is a way to understand the mistakes of the past and highlight future issues in our world. Photography is the storytelling of humankind. Images are a universal language that speaks to all of us, no matter our tradition, religion and language. I do not see animal photography without science and a conservation meaning. We're driven by an invisible force deep within us, that's why we photograph animals.

(above) Polar bear crossing melting ice, Nunavut, Canada, 2017
(overleaf) Crabeater seals resting on ice floes after feeding, Antarctica, 2018

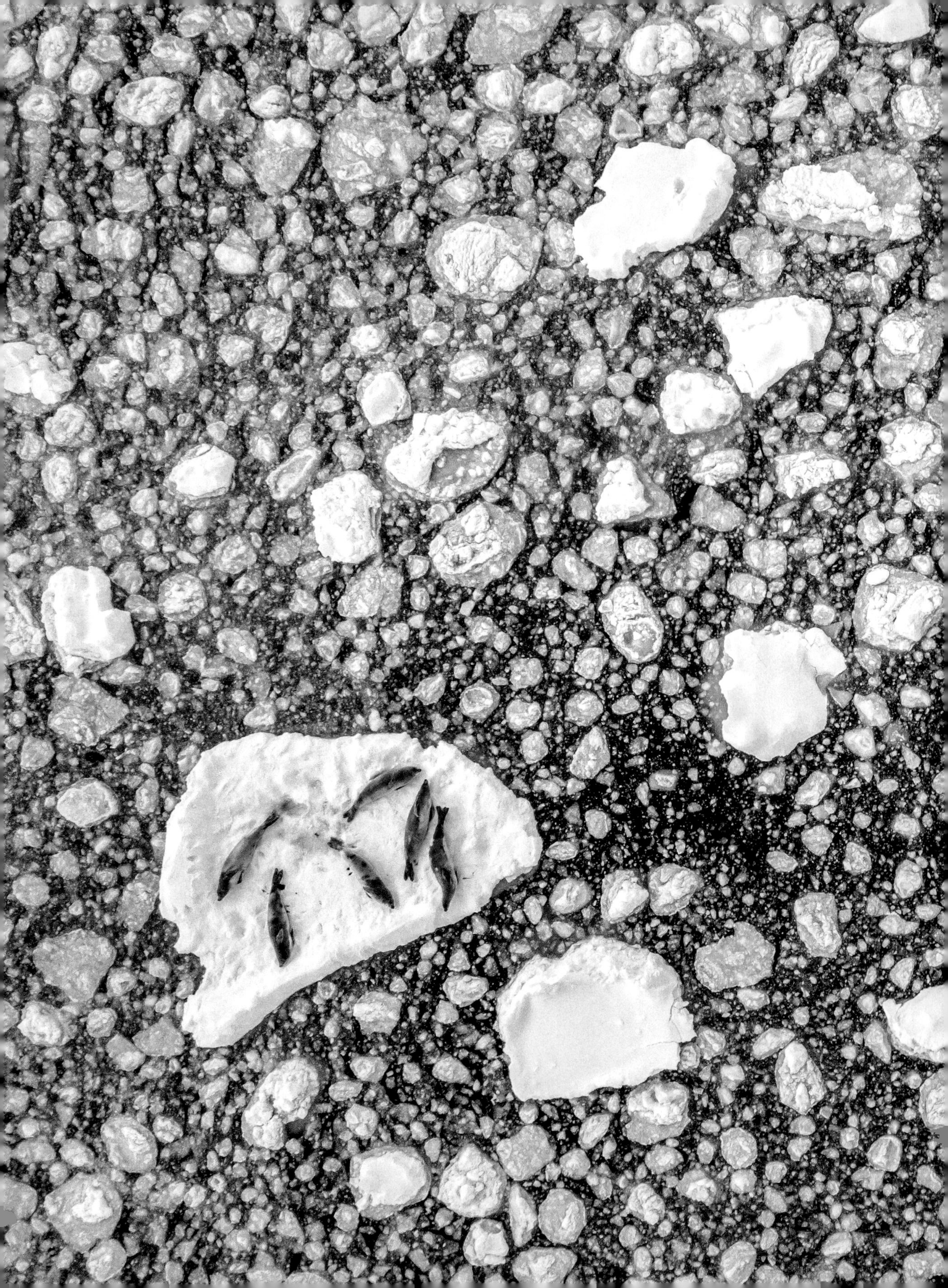

Walrus colony haul-out, Svalbard, 2020

Ingo Arndt

INGO ARNDT *spent every single minute of his spare time as a child outdoors in nature and later plunged straight into the life of a professional photographer. His picture stories* Animal Feet *and* The Pumas of Patagonia *each received a World Press Photo award.*

Can you remember the first photo of an animal that you took? It was a kingfisher. I was a child, still very naive, and approached it with my camera in hand and a 135 mm lens. A bit later I offered it fish in front of a wooden self-made hide and was much more successful. I found the birds in a forest close to my home. I went there on my bike almost every day, no matter if it was snowing or the sun was burning.

Is there a particular photo, or body of work, that raised your curiosity like no other? Michael ['Nick'] Nichols' book *The Last Place on Earth* came out in 2005 and it combined artistic skills with modern gadgetry. New technology allowed him to set aside the usual telephoto lenses that magnify and flatten subjects, freeing him to use more normal lenses with special lighting. The pictures helped bridge the gap between wildlife and humanity better than ever before.

What do you hope to achieve? I'm very lucky to live in a world that still has a few 'intact' wild places. And I'm lucky to have the chance to visit many of these places as a wildlife photographer. I think it's my duty to document this and what future generations won't see anymore. It breaks my heart when I see what we humans destroy and how many places in the world have already changed that I know differently from trips some years ago. Of course, I want more than just to document: I hope my pictures also have an artistic quality and that as many people as possible enjoy what they see. Photography is a very useful tool in environmental protection. If I can show people the magnificence of nature and can help a little bit to protect it, I'm very happy.

Has an animal photo forced you to examine your life? Images of factory farming have moved me to rarely eat meat and, if I do eat it, I eat only game from the area where I live. It's a shame what we do to animals to satisfy our hunger for meat.

Can you describe the positive impact that one of your photos has had? My last assignment was on wild honeybees. I documented the behaviour of an animal species that we humans keep in large numbers as a honey producer. I think my pictures can stimulate beekeepers to rethink beekeeping.

It is often said that 'photography can change the world'. Do you agree? Absolutely I do, no question. It is only a small part but an important one. Used correctly, you can reach a lot of people with photography. It is an important tool for nature conservation.

Is there an animal photo that you see as your stand-out shot so far? The puma hunting a guanaco. Without this picture the story would not be complete, it is the 'key picture' that I needed to tell the whole story of its life, behaviour and environment.

Do you have any other thoughts to share? I keep hearing people saying 'nature will survive without us'. That's not wrong, but don't we have a duty to protect the animals and plants that live together with us on this planet? Do we have the right to exterminate countless species? How can a living being be so selfish and place itself above all other living beings as we do?

Puma hunting a guanaco, Torres del Paine, Patagonia, Chile, 2017

‘Photography is freezing the moment. It gives us the chance to pause, to concentrate on a moment in a world where everything is moving, everything is getting faster and faster, where everything disappears so quickly.’

Ingo Arndt

(opposite) Honeybees colonize a black woodpecker nest cavity, Germany, 2019
(above) Thick-tailed scorpion mother carrying her young, photographed under controlled conditions in a German collection, 2008

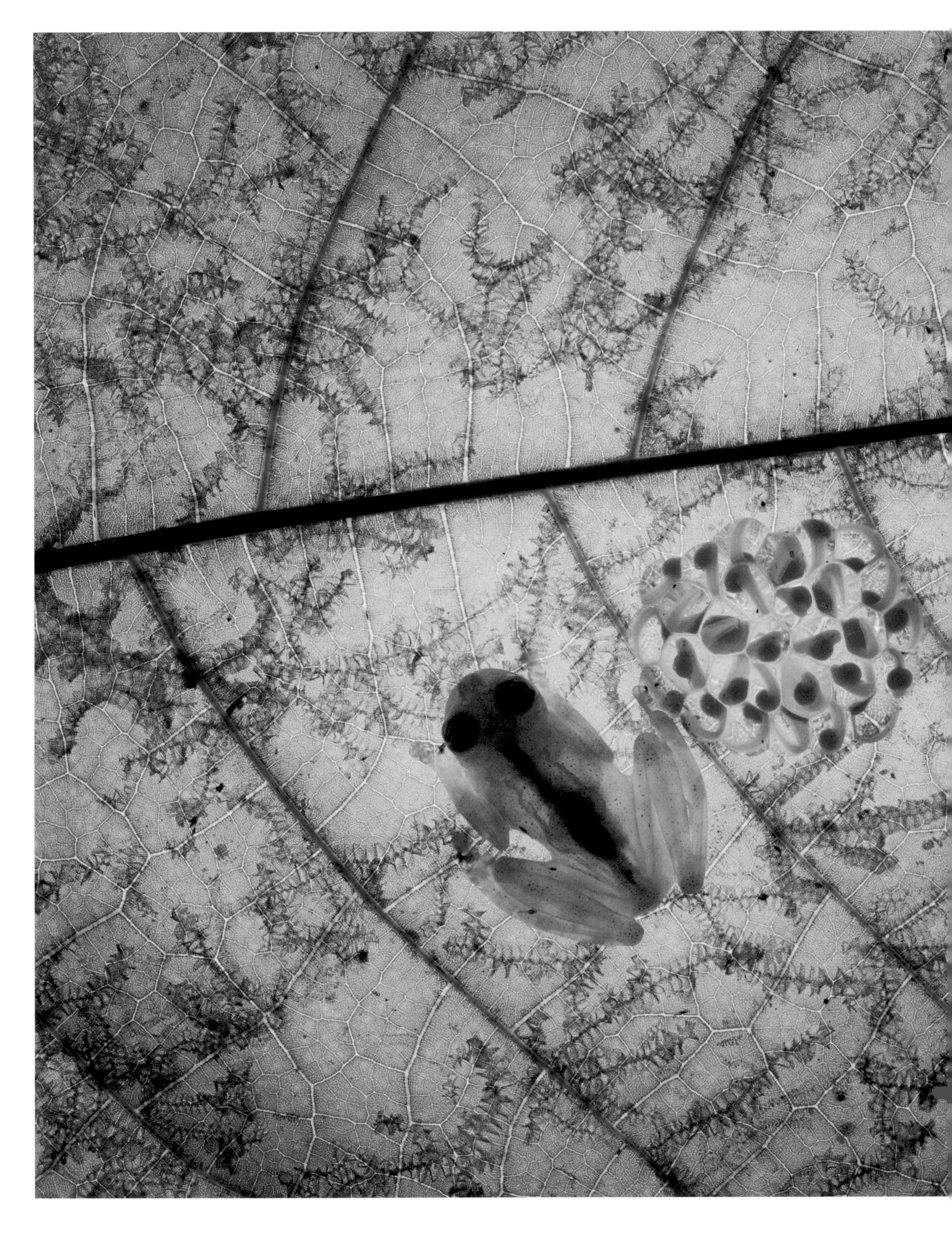

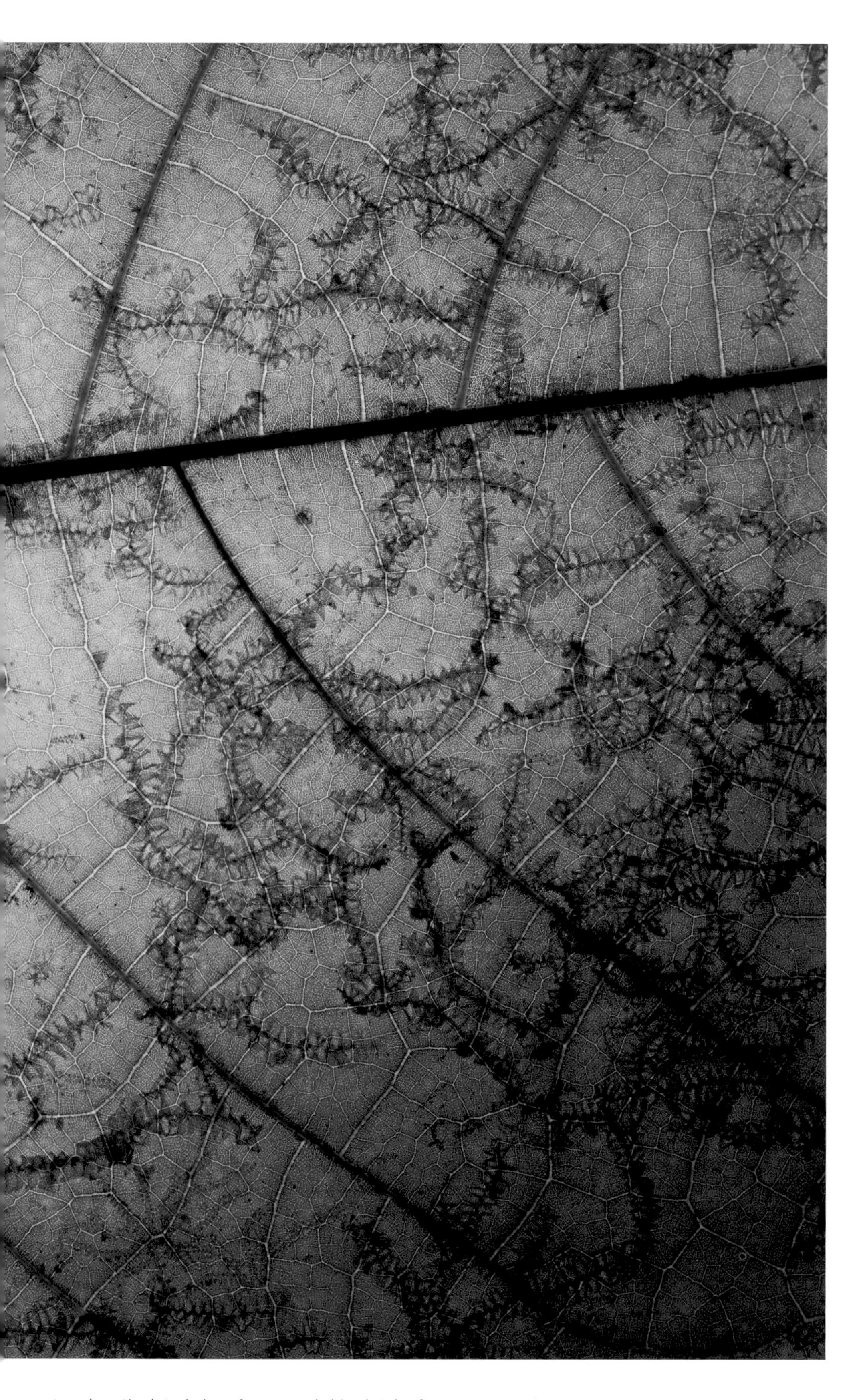

A male reticulated glass frog guards his clutch of eggs, Costa Rica, 2012

Traer Scott

TRAER SCOTT *is a fine art and commercial photographer and the author of fourteen books. Her work is exhibited around the world and has been featured in* National Geographic, LIFE, Vogue, *the* New York Times *and dozens of other print and online publications.*

Is there one historic shot that you really admire? I am a little obsessed with the famous photos of the thylacine, the 'Tasmanian tiger', from 1930. It's terribly sad and haunting to see it in captivity and to know it's the very last of such an extraordinary, alien-looking species.

Is there a particular photo, or body of work, that raised your curiosity like no other? I adore Ami Vitale's work with pandas. It is, in my opinion, one of the finest photo essays on the modern human–animal relationship ever created.

So, why do you photograph animals? I like to say that animals make me human. I may or may not have been the first to coin that phrase, but it's the best way I can explain such an indispensable part of me. Since animal rights and welfare have always been a huge part of my life, I think I was destined to end up focusing my career on animals. It took a while for me to unlearn the prejudices of the art world and wade through the tropes that we are presented about animals, but then I realized the power and influence that the right project can wield.

What do you hope to achieve? I pretty much always have the same goal, which is to elevate the perception and opinion of non-human animals; to inspire empathy and compassion. I'm particularly drawn to portraiture in order to celebrate what and *who* animals are, not in an anthropomorphic way – I think that comparing animals to humans actually belittles them – but as a way to celebrate their own unique abilities, emotional depth and intelligence.

Has an animal photo forced you to examine your life? I often think about Chris Jordan's albatross images. It's hard to look at them and not examine your use of plastics and how irresponsible we are with how and where we discard them, especially the smaller bits that are so easy to overlook. This photo inspired me to buy less single-use plastic, but I am angry and frustrated by the lack of plastic-free choices for consumers.

Is there an animal photo that you see as your stand-out shot so far? This is hard and I might be way off, but I feel that Bailee from *Shelter Dogs* might have that distinction. It is an image of a soulful dog who was euthanized needlessly, just to make space in a shelter. The simple caption used in that book – 'Bailee, Euthanized' – and the image together are extremely powerful I think. I have plenty of photos of dogs who didn't make it, and they all affect me, but even after all these years this one still takes my breath away and reminds me of why I started making images.

Are animal photos important? Meaningful animal photos are exceptionally important because they generate empathy and action, unlike the trite meme-like content that we are constantly inundated with. I feel that animals as subjects, serious subjects, have only come to be accepted in the past twenty years (outside of traditional wildlife photography). Boxing non-human animals into cutesy tropes is, among other things, a way to marginalize their lives and distract from the suffering that we inflict on them. It's much easier to kill or exploit a creature when it's constantly reduced to stereotypes in the media.

Bailee, a pit bull-Great Dane mix, picked up as a stray in Rhode Island, 2005

(opposite, above) Gazelle, American Museum of Natural History, New York, 2009
(opposite, below) Tiger, Academy of Natural Sciences, Philadelphia, Pennsylvania, 2010
(above) Bison, Yale Peabody Museum, New Haven, Connecticut, 2014

Kate Kirkwood

KATE KIRKWOOD *is a photographer working from a farm base in the Lake District in England. She is especially drawn to hunting the dynamic moment, a happenstance, the serendipitous, in both rural and urban settings.*

Who are the wildlife photographers that you most respect? I enjoy work by photographers who succeed not because of expensive kit, complicated journeys and HDR, but who are either thinking beyond the stereotype of the hunter-photographer, or who are alert to serendipitous encounters in their everyday lives. Stephen Gill, for example, whose poetic work includes *Please Notify the Sun* and *The Pillar*. He's perhaps an artist rather than a wildlife photographer, but why should these things be separate?

Do you have a favourite animal photo that you have taken? One of my series of cow spines, with a rainbow at dawn. Also, a ewe and lamb trotting down the road towards a sliver of crimson sky ahead of the car in the last light of day. A third image, if I can have one more, is a flash shot I took while talking on the phone and happened to see out of my window. I walked out, still on the phone, and snapped the sheep under the big sycamore at dusk. The tree, rocks and rain make beautiful patterns around the mysterious, bright eyes of the sheep. Pictures that end up really resonating emerge at all kinds of moments and often when you're not searching for them.

Is there a particular photo, or body of work, that raised your curiosity like no other? Like many, I was hugely impressed by Sebastião Salgado's *Genesis*, but I also felt uncomfortable about it: the expense of his expeditions, funding, teams working for him. And, despite his huge compassion and intelligence, the incredible sheen and perfection of his shots, the epic quality of the project, verges on the sentimental in my view. I find perfect images that are impossible to see with the naked eye or by average humans are awesome, but distancing. So *Genesis* really aroused my curiosity because I wondered how raising awareness of the damage and destruction of the Anthropocene might be done in other ways. Simpler, less grand, less aggrandizing. More local.

Can you describe an animal photo that you would most like to capture? If it were not for the damage of international travel, I would love to document the close relationship many communities still have with cows and the hugely central role the animals have for these families. It's easy for us in the richer world to choose to no longer consume meat or dairy because we are so alienated from the sources of these products. Reliance and tradition affect others differently.

So, why do you photograph animals? Initially, because they are sometimes available and willing. Well, *sort of* willing – domestic animals are. Because I live a fairly isolated life, the creatures I find living around me were the easiest, cheapest and most interesting element of the landscape to explore when we bought the first family digital camera. But I suppose that answer doesn't get to the heart of *why* I do it. I turn to the animals around me, in the hope of capturing – sharing and raising awareness of – their sentience and the complexity of their communities.

Can you describe the positive impact that one of your photos has had? I guess any photo of mine that has delighted a viewer in a small way: something they recognize, or something they're seeing anew that pleases them, usually one which celebrates animals in some way. A delicate blue tit at a feeder, outlined against a wet window, is one.

Are animal photos important? Photographs of animals reminds us who *we* are: our gentleness, power, cruelty, negligence, love, care.

Cow spine with rainbow, Cumbria, England, 2010

‘Photography is driven by who we are and what we prioritize. For me it’s a febrile constant of being alert and open to possibility, to my environment, the weather, to the intense magic of a living world when it brushes briefly against me.’

Kate Kirkwood

(opposite) Cock on the Hawes Farm track, Cumbria, England, 2010
(above) Ewe and lamb on the fell road, Cumbria, England, 2012

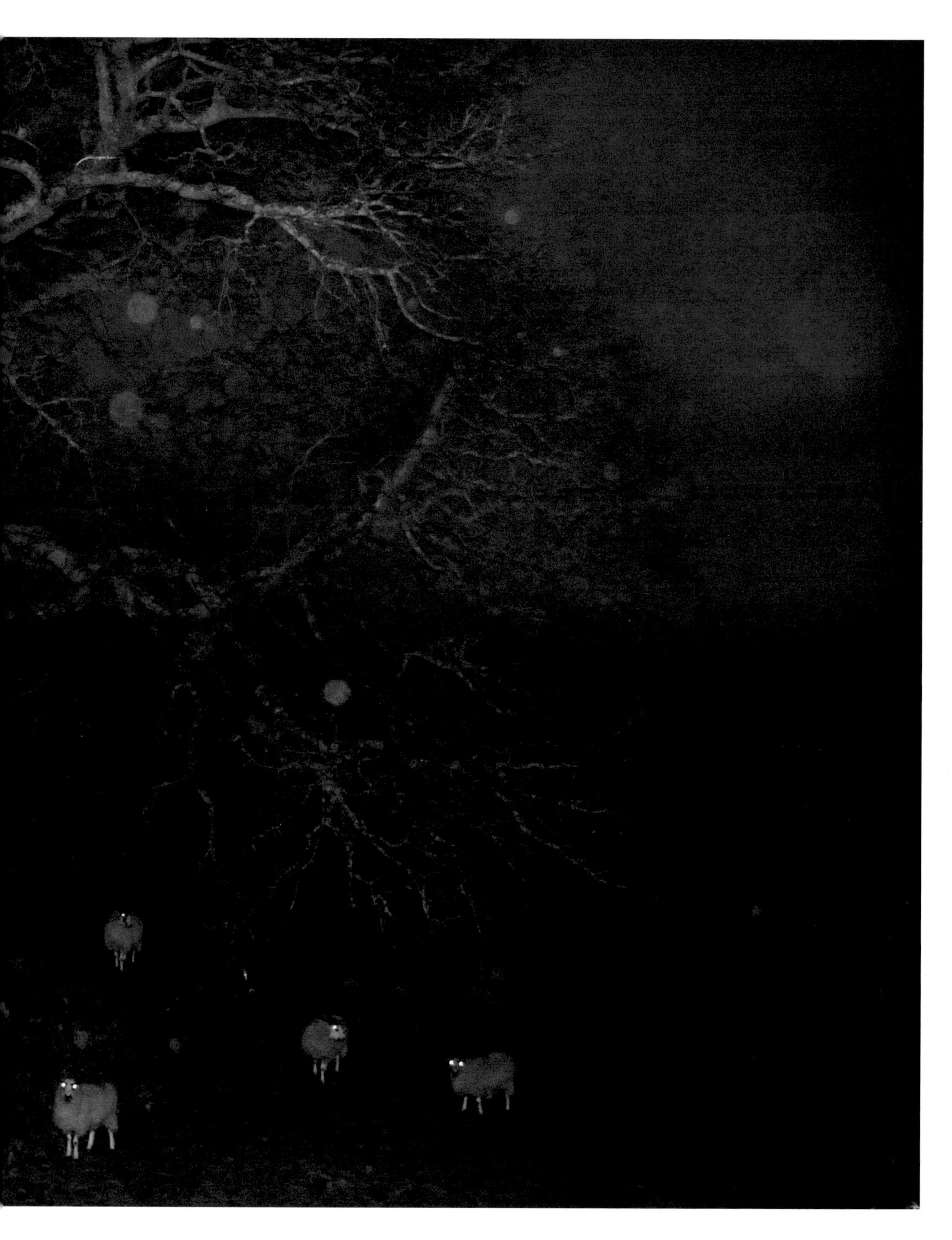

Sheep under a sycamore, Cumbria, England, 2011

Jo-Anne McArthur

JO-ANNE MCARTHUR is a photojournalist and animal rights campaigner who documents our complex relationship with animals around the globe. Since 1998, her work has taken her to more than sixty countries and her images have been used by hundreds of organizations in their advocacy efforts.

Can you remember the first photo of an animal that you took? I think the first photo of a non-human animal I took that actually mattered – because it was an attempt at an outward expression of how I viewed our relationship with animals – was of a donkey at Papanack zoo near Ottawa. It was one of the first pictures that expressed my sorrow about how we treat others.

Describe an animal photo that you've missed or not taken: I think I've got the images I have set out to take. Many have been far from perfect but have managed to document how we treat other animals. They are proof and so many of my imperfect images have been useful to educators, campaigners and journalists.

Have you ever risked your life for a photo? Unfortunately that's part of the deal when you do investigative work, be it covering animal industries or more traditional conflict photography. With both, the real danger is other humans. I'm seen as a threat to people's livelihoods and to animal industries, so my colleagues and I are always at risk of beatings, fines, jail time and being charged under the ever-looming 'ag-gag' laws and, in the USA, the Animal Enterprise Terrorism Act.

Who are the photographers that you most respect? Street photographers and war photographers informed my animal work. The lengths to which they go to tell a story that they passionately want to document and share always inspired me. How to take images of the horrors we inflict on one another in a way that captures the viewer, awakens their empathy? Sebastião Salgado, James Nachtwey, Larry Towell – I watched what they did and ventured to do the same for human-animal conflict.

So, why do you photograph animals? Important issues in our world need visibility as one of the first steps to getting people to see, understand, care and change. That's why I love contributing to the canon of photojournalism. My team and I get to play a small role in sharing and changing the course of things. Humans cause so much suffering to other animals and this needs to be uncovered, documented and shared. So the *why* here is important. It's not just a job for me. It's a life mission.

It is often said that 'photography can change the world'. Do you agree? Yes, I agree. That's the reason I do it. Photos are a catalyst and people must then be the change. I have hope in human compassion.

Is there an animal photo that you see as your stand-out shot so far? There are some shots that you just know are special in that fraction of a second that it takes for you to make them. When I saw the kangaroo in the burned eucalyptus plantation, she was a few hundred feet away from me, quite some way from where I knew I needed to be to get the shot. That was a long walk! I prepped my camera to the settings I wanted and walked calmly as she watched. Got to where I needed to be: where the rows of burned trees faded symmetrically into the background. Click. Crouched down. Click.

What does photography mean to you? Photography is how I express my feelings and ideas about the world. It's how I help. It's how I bear witness and convey the joys and sorrows of the world, which are also my own. Photography is my passport into the lives of others and a way to both satisfy and amplify my curiosity about the world.

Eastern grey kangaroo and joey in a burned eucalyptus plantation, Mallacoota, Victoria, Australia, 2020

‘I document animals, but truly these photos are a sociological study of *we* animals. Inevitably, my images reflect how varied are our reactions to other animals: fascination, indifference, amusement, boredom, fear, even awe.’

Jo-Anne McArthur

Shots from the series *Captive*
(opposite) Jaguar, France, 2016
(above) Baltic grey seal, Lithuania, 2016
(overleaf) Children with orangutans, Denmark, 2016

INSIGHT

Jim Naughten

Wildlife Fictions

Jim Naughten *is an artist exploring natural history using photography, stereoscopy and painting. Naughten's work has been widely featured in exhibitions across Europe and the USA, including solo shows at the Imperial War Museum, Horniman Museum and the Wellcome Collection.*

In the evolutionary blink of an eye, humans have come to dominate and overwhelm the planet. I'm interested in how far our relationship with nature has fundamentally and dangerously shifted from that of our ancestors. I create digital artworks to conjure worlds that feel familiar yet strange; photography and painting combined. From orangutans swinging through psychedelic vegetation to grizzly bears or gorillas emerging from the depths of mysterious forests, my work explores the idea of the natural world as a faraway fictional fantasy. I want to alert us to its rapid disappearance and our growing estrangement.

I began my journey as a painter but switched to photography at art school, partly because the cool kids were doing it but also because I imagined pressing a button would be much easier than agonizing over an empty canvas. That did not play out quite as well as I hoped. Many years of carting heavy equipment around, dealing with troublesome sitters and a brief spell in an African prison put paid to the fantasy of simplicity – and that was one of the easier photographic projects! Some shoots were just intolerable. Humans are much more savage and frightening than any wild animal you might imagine.

I slipped into the murky world of advertising and magazine photography and got stuck there for years – my wilderness years! I was never particularly happy but it took me a long time to realize that I was no longer making pictures for myself. Before long, I was desperate to find a subject that was closer to my heart. It took a few years for it to dawn on me but I finally found it: animals.

Well, to be specific, wildlife. But wildlife that was no longer roaming about in wild areas. I found my subject in specimens. While working on a separate project, I stumbled across stereoscopy, an early form of three-dimensional viewing technology that's even older than photography. I absolutely love it. It's accessible, fun and can give the subject an extraordinary feeling of presence. It's a little awkward in the sense that you have to physically engage with it – one viewer at a time, in a gallery or with a book and a stereo viewer – but that is what gives it a special power. It is both charming and immersive. It has directness, albeit in what seems an old-fashioned way.

My first foray into stereoscopy resulted in *Animal Kingdom*, a book and several exhibitions. When I was young I adored dinosaurs, fossils and anything to do with natural history, so it was wonderful reconnecting with this childhood obsession. I started work at the Grant Museum of Zoology in London and began making stereoscopic images of articulated primate skeletons. They worked exquisitely in three dimensions and one particular chimpanzee skeleton looked so utterly human, that it reminded me that we are primates and very much part of the natural world, despite believing that we are different.

Working in natural history museums has been a wonderful privilege. They feel like

Gorilla, from the series *Mountains of Kong*, 2019

Stereoscopic images from the series *Mountains of Kong*, 2019

‘Photography is a storytelling superpower. Work can create awareness and discourse about our disconnection from the natural world, our fictionalized ideas about nature and possibilities for positive change.’

cathedrals and temples to the natural world. Working in relative silence outside museum opening hours, I can feel the history behind the collections, in the presence of scientists from Darwin’s time up to the present day. The specimens are innately fascinating and often extraordinarily beautiful, but there is a paradox because the animals did not die a natural death and have been hung, drawn and quartered, stuffed, articulated or preserved for human observation for eternity.

There is a wealth of subject matter in these collections. My next project, *Mountains of Kong*, came about after hearing someone talking about the eponymous mountains on the radio. I loved the name and the story – essentially a mountain range across Central Africa that people in Europe believed existed for a hundred or so years, before discovering that it was never there – and I went back to a well-learned lesson from college. One of our tutors suggested we imagine an empty gallery and picture what we want to see on the walls. It’s a simple idea but it helps clarify the idea that if you could photograph anything in the world, what would it be? At that moment, obsessed with stereoscopy, I decided I would go back in time to a place that may or may not have existed to make ‘scientific’ stereoscopic images of its fauna and flora. The creatures of Kong emerged from this simple idea.

This time I worked with dioramas in natural history museums in the US and Europe. The process of making the stereo images is quiet and meditative. I think of these museums as sacred memorials to the natural world, which provides another layer of sadness I suppose. When I’m working it’s just me and the ghosts of the poor creatures behind the glass. I make the stereoscopic images with a single camera on a slider that is mounted on a tripod, with one image for the left eye and one for the right eye. Making the photographs is relatively straightforward; exhibiting them or showing them is when things get tricky.

I found a way of making viewers with mirrors that allow large-scale print viewing, alongside more traditional ‘in-line’ stereoscopes, which are my modern versions of the Victorian originals. It’s a restrictive way of showing work in this modern age, I appreciate: one viewer at a time. It means they can only really be seen properly in person. It’s not something you can put in a magazine or website, but I enjoy creating within these limitations.

While I was working on *Kong* I visited the Field Museum in Chicago. It was absolutely full to capacity and hard to move around but there was one room that was completely empty: an exhibition on extinction. The first text stated that 30,000 species are going extinct every year because of human activity. There was a digital display showing that since eight o’clock that morning twenty species had gone extinct – and it kept ticking away. Four species every hour. Over eighty each day. Maybe more. It was deeply depressing on two counts: both the rate of extinction and also the lack of interest from the people at the museum that day.

I decided then and there that my next project had to address this. The images would have to be engaging and eye-catching, but above all not off-putting for the viewer. My idea was that our rose-tinted view of the natural world was a largely fictional one. Idealized. Using taxidermy animals again, my images would explore these illusions and our disconnection

Mule deer, from the series *Eremozoic*, 2021

from the natural world. Dioramas are often my starting point. Human-made fictions. Old windows on to the natural world, then with elements added and subtracted, colours altered and abstracted.

The celebrated biologist E. O. Wilson suggests we are entering the 'Eremozoic' period – following the Palaeozoic, Mesozoic and Cenozoic periods bringing us to the present day – categorized as the age of loneliness or desolation. For the first time in 542 million years of animal life on earth, a single species is responsible for the decline and extinction of the majority of other species on the planet.

By creating unreal settings I'm suggesting that our image of the natural world is fictional. The images are designed to be surreal and utopian but absolutely not representative of reality. The future does have the potential to be extremely dystopian but it seems that awareness of the gravity of the situation is growing dramatically, so I am not giving up hope just yet. I'm also interested in how we have become mentally disconnected and insulated from wild animals. Wildlife happens elsewhere: in isolation, on nature documentaries, safari parks and zoos. It feels like a secondary consideration as humans (and human activities like agriculture) increasingly dominate every corner of the world. There is yet another fiction at play here, because the reality is that we are still a biological species and still subject to the laws of nature, still susceptible to viruses, still just a few missteps away from extinction ourselves.

Art can make a difference. Not perhaps by just hanging in a gallery, but ripples of thought can spread fast. We have no time to lose, so all kinds of creative activity matter, however small, however local, however strange. Though it might not seem obvious in my fields of pink grasses, jars of specimens or visions of beastly creatures, even odd things like this can begin a conversation that can lead to positive change much further afield. Out in the real world.
In whatever shape, size or style, photographs of animals matter.

(opposite) Brown bear, from the series *Eremozoic*, 2021
(overleaf) Leafy seadragon, from the series *Animal Kingdom*, 2016

Daniel Naudé

DANIEL NAUDÉ *is a South African artist who uses photography to explore human stories in the environment. Naudé's first book,* Animal Farm, *was published in 2012, followed by* Sightings of the Sacred *in 2016. His work* The Bovine Prophecy *was shown in London in 2022.*

Can you remember the first photo of an animal that you took? It was our dogs at home, taken with my parents' Nikon film camera. I think the early years of doing this definitely had an effect on my professional life photographing Africanis dogs. Whether pets, or livestock or animals fed in the garden, we like to be close to them; to surround ourselves with the thought of animals, if not always with the reality of animals and their needs.

Is there one historic shot that you really admire? The masterwork of Eadweard Muybridge, specifically his study of motion in animals. I find it so interesting, energetic and *exploratory*. That's the thing: working out the technical limitations and then overcoming them. Extending the photographic possibility. And all the while, showing us something new, hidden in plain sight: the movement of animals.

So, why do you photograph animals? Because I am *interested*. Humans have built their lives and ways of living around animals since the beginning of history. The need for the wild and uncertainty has always been, and will stay, in the instinct of human beings, no matter how developed our civilizations become. I find the fine line between domesticated, feral and wild animals fascinating.

It is often said that 'photography can change the world'. Do you agree? Absolutely. Because a visual image breaks through language and is readable in some way for all viewers. That is the reason why my work is based on animals. I don't focus on a subject, an event or certain groups of people but on animals that everyone can relate to.

Is there an animal photo that you see as your stand-out shot so far? A dog portrait: '*Africanis 12*, Richmond, Northern Cape, South Africa'. What's fascinating about these dogs is how mentally and physically adapted they are to the areas where you find them. Animal form changes with geography, so you get Africanis dogs in dry areas that are more slender and have shorter hair, but in coastal areas they are larger in size with thicker coats. Strength and diversity – it seems very South African to me. A mix of cultures and mixed identities. It took me ages to edit. I would just print everything out, scattered on my apartment floor, and agonize for days. How can I start? In the end, the simple images won out and my style has stayed in that place ever since. I'm searching for stillness in my work, that's the foundation. Though I'm moving hundreds of miles around our country to meet farmers and find the wildlife I'm interested in, it's all about stillness for me. I want to capture the *spirit* of the animal.

Do you have any other thoughts to share? A single image can change the way people appreciate a species. For sure. Digital media and the Internet have exposed us to a global audience. Anyone can post across social media and write something about an image. All kinds of personal reporting and news platforms too. This gives photographers the potential for a voice like never before.

Africanis 12, Richmond, Northern Cape, South Africa, 2009

Nguni bull, Kei river, Eastern Cape, South Africa, 2009

Quagga, Stellenbosch, Western Cape, South Africa, 2010

Nguni cattle farmer Ben Fyfer at his desk in Louwna, North West Province, South Africa, 2010

Mario Jacobs with an African clawless otter, Quaggasfontein farm, Eastern Cape, South Africa, 2010

Georgina Steytler

GEORGINA STEYTLER *is a self-taught nature photographer with a passion for birds, ethics and conservation. In 2018 she became the first Australian woman to win a category in the Wildlife Photographer of the Year.*

Describe an animal photo you've taken that was memorable to you. A Buller's albatross. I took it in rough seas from a small boat – I was so seasick I'm sure my vomit was bringing the albatross in and it was raining – yet miraculously I got a sharp image of the most beautiful animal staring back at me.

Do you have a favourite animal photo that you have taken? Any leaping mudskipper photo – they just make me so happy. I have always liked them but an old birding friend told me the best location to get mudskippers up to a foot long and when I went there I was immediately in love. They are the funniest little critters, jumping and arguing and patrolling their territories all day long. I have since done quite a bit of research about them and they are remarkable examples of the evolutionary transition of life from aquatic to terrestrial, with eyes on top of their heads (so they can see predators above the water) and an ability to tolerate low oxygen levels. Even though they are not endangered, I still think it's important to document these kinds of animals as it helps people to appreciate all kinds of nature, not just the big, cute and cuddly mammals.

Is there a particular photo, or body of work, that raised your curiosity like no other? I could not get over the image by Frank Deschandol in the 2019 Wildlife Photographer of the Year awards of a weevil that had been taken over by a 'zombie fungus'. For a long time afterwards, I told every person I met about how amazing it was and made them look at the image. I still cannot get over how a fungus can invade the animal and influence it to climb up a tree. Then it grows stalks from the dead animal's head, ultimately releasing spores into the world below. It seems like something from science fiction, but it's incredible and real. It reminds me that we humans have so, so much more to learn about nature and the world around us.

So, why do you photograph animals? I much prefer animals to humans. I suffer from depression and anxiety and, unlike humans, animals relax me. I'm never happier than when in wild environments with an animal, watching them do what they do naturally.

Can you describe the positive impact that one of your photos has had? In 2018 my image of two mud-dauber wasps won the invertebrates category of Wildlife Photographer of the Year. A month or so after the awards, I received a call from a university professor who was dating Aboriginal rock art in remote Australia by carbon dating the ancient mud-dauber wasp nests over which some of the paintings were made. He had seen my image and wanted to learn more about the behaviour of the wasps that I had observed. When I took that photo, I never dreamed of the kinds of connections it could make.

Are animal photos important? The greatest threat to our wildlife is the disconnect of people from nature in the modern world. How do we reconnect? By going beyond conventional shots, or testosterone-fuelled hunter-and-hunted nature documentaries. Studies have shown that utilizing anthropomorphic language when describing dogs creates a greater willingness to help them in situations of distress and that the degree to which individuals perceive sentience in other animals predicts the moral concern afforded to them. Photography can directly affect animal welfare, right now.

Buller's albatross, Indian Ocean, 2015

Mudskipper, Roebuck Bay, Western Australia, 2015

Mud-dauber wasps gathering material to build their nest, Western Australia, 2016

Mateusz Piesiak

MATEUSZ PIESIAK *is a wildlife photographer from Poland with a passion for birds. His work has been published in journals such as* National Geographic, BBC Wildlife Magazine *and* The Guardian. *He has been awarded prizes in numerous competitions, including Wildlife Photographer of the Year.*

Is there a particular photo, or group of images, that inspired you into action? I especially remember the wide-angle series of photos by Bence Máté showcasing pelicans. For me it opened up a new way of photography. After seeing Jan van der Greef's image of herring gulls I started experimenting with long shutter speeds. And Jasper Doest's flamingo series really made me realize how important the story and the message in a picture are. I once asked Jasper what lens he took these photos with. He replied that it didn't matter, he could take it with any gear, what is important is what you want to tell through your image and how. It stuck in my mind.

So, why do you photograph animals? I photograph animals because I am able to capture moments that are elusive to the human eye. Often the action happens so fast that it's hard to see anything in real life. Thanks to modern techniques, cameras are able to take a lot of frames per second, so there is a good chance of capturing that perfect wing or beak pose. I'm also working with hospitals in Poland and Germany: my works decorate the walls and make patients' days brighter. I guess that's also another big *why* – to put a smile on people's faces and help them. I know when I create images, there is a big smile on my face too.

Can you describe the positive impact that one of your photos has had? I received a message from a lady who had been struggling with depression for a long time. One of my photos gave her a lot of positive energy. It was a shot of a cygnet. She was in a difficult mental period, and through my simple image she was able to see the positive side of things again and appreciate the world around her. I get many messages from people who are starting to notice the different species of animals that live around them and want to care about them (and nature as a whole). It might sound like a small thing, and I suppose it is, but it's also fundamental. It's a major reason why I shoot. Opening people's eyes to small things, awakening minds to beauty worth protecting.

It is often said that 'photography can change the world'. Do you agree? A single photo is unlikely to change the world, but in combination with text photography can be a powerful tool in sharing the message. I believe that an image can convey much more information and emotion than text alone. Think of issues like poaching elephants for tusks or hunting sharks for fins. Good images evoke very strong emotions and are remembered. People are in a constant rush and can't focus their attention for too long. They often read the text very cursorily; a strong photo can effectively catch the eye. Properly used, preferably in a wider photographic project, it may not change the whole world, but it will certainly affect some aspect of its perception.

Are animal photos important? In Poland, we have a proverb: 'what the eye does not see, the heart does not grieve over.' I believe that if there were no photos of wild animals showing their beauty and diversity, most people wouldn't even be aware of their existence and certainly wouldn't spend money on their protection. When the need to protect nature is so important and we are losing irrevocably more species day by day, good images and honest stories are vital. That's why I believe that animal photos are important and can influence the way we perceive the world. They can encourage us to take care of our planet.

European bison, Białowieża National Park, Poland, 2017

(above) Egrets and gulls, Poland, 2014
(overleaf) Bramblings in a field of old sunflowers, Dolny Śląsk, Poland, 2021

Karim Iliya

KARIM ILIYA is a photographer, drone pilot film-maker and whale-swimming guide based in Hawaii and Iceland. He aims to use his photography and video to share his unique perspective on the earth while telling stories about people and animals.

Can you remember the first photo of an animal that you took? One of the first photos I was really happy with was of a baby anole clinging to the stem of a flower. I have a print of it on my wall. Even now, over a decade later, it's still in my portfolio.

Who are the wildlife photographers that you most respect? I have a deep respect for everyone who paved the way and brought us the stories that have helped people fall in love with and protect animals. This includes all the camera crew from BBC series like *Planet Earth* and *Blue Planet*.

What is the most influential animal photo ever taken? Any of the old photos of animals that showed people a new animal for the first time. In modern times, probably Ami Vitale's work with rhino has had the most influence. Compassion, craft and simplicity of shot. Her images are seen by millions of people each year.

So, why do you photograph animals? We often pile animals into one category and humans into another, but the truth is we are just one of the millions of animals on this planet. Humans are fascinating and I love to photograph them as well but the diversity within the natural world is incredible. We have animals like sperm whales that battle with colossal squid, seemingly using superpowers like sonar to stun their food. There are peregrine falcons that hunt at speeds faster than a race car, animals like the octopus that can change shape and colour, coral colonies that compete overnight and animals that glow in the dark. There are countless animals that have evolved to be highly specialized to their environment. This also means that many of the most impressive species are in danger. As a species, we humans have taken more and more land and resources, poisoning our waters and soils without any regard for the creatures that live there. Photography can be a means of showing people the beauty of our planet before we lose these incredible species.

Has an animal photo forced you to examine your life? There's a photo by Jonas Bendiksen I saw long ago. It's not a conventional wildlife photo, in that the main subject is of villagers collecting scrap metal from a crashed rocket in a field full of hundreds of white butterflies. It reminds me that despite everything we do, life still thrives on this planet.

Is there an animal photo that you see as your stand-out shot so far? So far, for me, it's my photo of a humpback whale calf's eye. This animal was incredibly playful and would come over while its mother slept a little deeper. To me this is as intimate a connection you can get with a wild animal. When I photograph whales, I am typically not looking through my camera and just relying on my instinct so I can use my eyes.

Are animal photos important? Most people have limited interaction with wildlife. We rely on shows like *Planet Earth* and wildlife photos to open up a window into the lives of animals. It takes immense amounts of time and patience to see and photograph many of the threatened animals in our world. It's important for conservation, appreciation and for scientists to understand how our ecosystems interact and depend on each other. We're lucky to live in a time where we can create intimate images of animals so they are no longer seen as monsters to be feared. Animals are an important part of this world and if we forget that, we risk losing them. Without animals – bugs included – our ecosystems will collapse and our species will go extinct.

The eye of a humpback whale calf, Vava'u, Tonga, 2018

‘Photography is a way to tell stories and give voice to those who cannot, be it plant, animal or human.’

Karim Iliya

(opposite) Humpback whale calf breaching off the coast of Moorea, French Polynesia, 2020
(above) Hawaiian green sea turtle, Hawaii, 2021

Sea lion colony, Baja, Mexico, 2020

INSIGHT

Britta Jaschinski

Rethinking Things

BRITTA JASCHINSKI *is devoted to documenting the fractured existence of wildlife. She has been named GDT European Wildlife Photographer of the Year twice and is a category winner at the Wildlife Photographer of the Year competition.*

I USE MY CAMERA as a weapon to reclaim our essential respect for animals. I pick up my camera to fight battles, using photography as a catalyst for change. But even if you took photography away from me, I would strive to continue with the same mission.

My passion for animals and nature has always been there; I have never felt superior as a human. Because I had so much love for animals, for wildlife and nature, when I became a professional photographer – and that happened very early because I went straight into an apprenticeship and by twenty I was already working – I thought it made sense to combine my skills to create a powerful message for the world. It's something I now tell students when they ask how to become an animal photographer: is there anything you want to say?

From about the age of fourteen I was out collecting signatures, doing my bit to stand up for causes. I raised awareness about the cruelty behind the use of animals in labs for cosmetics and medicine and protested against nuclear power. When I moved to England to study photography, I experienced a feeling of home coming. Meeting like-minded people energized me and I was inspired by all the photographers I learned about. Some of them I even got to meet, as my college put a lot of effort into students getting experience in the 'real' photo-world.

My first project was about zoos – about the legal wildlife trade. Then I turned my lens to the illegal trade. Learning that the illegal wildlife trade is the fourth largest illegal trade – behind drugs, people smuggling and counterfeiting – worth an estimated £15 billion ($18.7 billion) annually, blew me away. Hundreds of millions of plant and animal specimens are traded internationally each year. I decided to try to capture that.

Back then I spent a lot of time in the darkroom, processing film and learning various printing techniques, which are still useful to me now when I process digital files, while also learning a lot about myself. Being in the dark, creating and looking at my photographic prints was almost like a window into my soul. How do we dare treat sentient beings like this? Thousands of animals across the globe endure severe physical and mental suffering in the name of entertainment, status, power, greed and superstition.

People say that my work has been instrumental in bringing change – in laws, in law enforcement, but also for the public across the globe. At a minimum I hope my images encourage more and more people to protect animals and our wild spaces.

I believe my photography has come a long way since those zoo shots in the 1990s. I've worked to show the horrific conditions in circuses and to expose the cruelty of animal training: electric shocks, pain, sleep, food deprivation, all kinds of horrors in the name of cheap thrills. I've searched out the grim trophies of the illegal wildlife trade, shelf upon shelf of beguiling yet terrible things seized at airports and border points. I want to understand

Captive beluga, from the series *Zoo*, New York, 1995

A zebra head confiscated on the US border and housed at the National Wildlife Property Repository in Denver, Colorado, 2016

Big cats, probably all hybrids of captive-bred animals,
Seven Star Park in Guilin, China, 2012

Leopard
Skins
Untanned
SHELF 9
SHELF 12
SHELF 13
SHELF 14

Leopard
Skins
Tanned
SHELF 2
SHELF 3
SHELF 4
SHELF 5
SHELF 6

why some people still demand these products even when it is pushing species to the brink of extinction. The problem is that the rarer the species, the more profitable they are when they are dead. People want that last rhino because it is worth so much more when it is dead. I want to be a voice for animals that cannot be heard. What drives me now is: what *needs* to be said?

Perhaps one of my most famous photos shows the head of a zebra that was confiscated by officers from the US Fish and Wildlife Service. I documented what gets confiscated and then held at a wildlife repository in Denver, Colorado. It is basically a huge warehouse containing millions of items. A shopping trolley is used to take items from A to B, for forensic testing and education. I felt it was ironic seeing animals in a shopping trolley and it added to my message: wildlife or commodity? It's like a superstore of human greed: relics, remains, trophies. Fragments of once wondrous animals. I feel ashamed being a human seeing these things.

Most people choose to document animals out in nature, but I've spent so much of my time looking into concrete cells, dirty cages and thinking about dead animals. The encroachment into nature continues. The illegal wildlife trade continues. Animal and human suffering continues. Needless death continues. My next big project is about virus spillover which, alongside climate change and biodiversity loss, is one of the most pressing issues of our times. It's like watching a slow-motion car crash right in front of our eyes. Why are we still refusing to do anything about it? Photography is vital here to illuminate the things that are hidden in plain sight.

I spoke to a scientist and he explained that every time a tree is chopped down in a rainforest and it falls to the ground there is an explosion of viruses. All the animals that once lived around the tree are immune. They have no problem, but of course humans enter the story, fell more trees and introduce their domestic pigs and other animals, and the viruses are amplified and enter the food chain. I learned that the chances of another global pandemic are high.

I am also creating still-life again: more animal objects – each item is a piece of the puzzle to see the bigger picture, to tell the bigger story; more confiscation contraband, products and dubious medicines. It's about trade and demand. Pathogens and pandemics. I am working with experts – biologists, special agents and other authorities. Sometimes you need a big team of experts to tell your story.

Once we raise awareness, we are halfway there – but then what next? How can we make sure people don't forget? That is why I love making books. It's the proof, the evidence and a reminder. If it wasn't for photographs the world's conscience would wither.

(previous pages) Leopard and tiger skins seized at US border entry points, 2016
(opposite) Turtle confiscated by customs officers and now held at the Leibniz Institute for the Analysis of Biodiversity Change, Hamburg, 2021

THE BONDS BETWEEN humans and other animals extend for thousands of years. Our ever-changing relationships might best be described as entangled: fraught with knots of contradiction and conflicting emotions. We love animals and yet we also destroy them. It's hard to think straight about animals. As a neat phrase in modern anthrozoology has it: 'some we love, some we hate, some we eat.'

Turn back for a second to take another look at the stereograph cards gathered at the beginning of this book: see there a writing spider, Bactrian camels, a pet horse, tourist dromedaries in Egypt, a little girl with her farmyard friends, exotic museum taxidermy, frozen Wisconsin fish, a just-shot Yellowstone elk, a hippo in Central Park Zoo, a giraffe in Antwerp, a cow, a dog, a caged bird, a prize racehorse, a horned lizard, a giant hippo called Caliph, a champion trotter named Rosalind, and a selection of Smithsonian stuffed birds. These stereographs – also called stereoviews – brought the wonders of the world into people's homes. Stereoscopy was a technology of vision that spread photography's reach as never before. Large companies commissioned photographers to travel to distant lands, publishing tens of thousands of views a day towards the end of the nineteenth century. This culture of looking, a kind of visual gluttony, remains in modern life: animal images are everywhere, even as animals themselves disappear from the natural world.

So, why do we photograph animals? Well, clearly there is no single reason. Intentions are varied and desires multilayered. It's complicated and often troubling. Photography is a tool of control and commerce, as much as it can be used with curiosity and compassion. The act of creating photographs in the past – in 'taking', 'capturing', 'shooting' – might be read as a process of trying to master nature through technology, while also seen as an expression of our care for animals. Sentiments shift through time and photographs reflect attitudes as much as shape them. Photography helps us learn more about animal behaviour just as we impose our fantasies upon them: is anthropomorphism, for example, still a guilty pleasure or can it be a critical tool to increase concern for our animal kin? What role might future photographic technologies, like advanced AI, play in complicating these relationships further? Where is this destruction of nature leading us? Is there any hope? These questions may well be beyond the scope of this book, and yet connections can be made between the photographic images gathered here, depending on how they are read. Think for each, why were they made? To what purpose were they put? How might their meanings change over time? How do these images make you feel? And, what about the animals themselves? In all of our activities, surely now we must be even more aware of our impacts on them.

John Gay, reference sheet, 1970

Louis-Jean Delton, woman on horseback, Paris, 1870

August 28, 1948

PICTURE

POST

HULTON'S NATIONAL WEEKLY

PHOTO-FINISH

HOW IT WORKS

AUGUST 28, 1948

Vol. 40. No. 9

4D

Race Finish Recording Company, photo-finish cover story, *Picture Post*, 1948

David Fairchild, king grasshopper, *Book of Monsters*, 1914

Thomas Marent, squeaking silkmoth, Qinling Mountains, China, 2009

Oliver Meckes, tardigrade, a micro-animal barely 0.5 mm when fully grown, 2009

Bruno D'Amicis, three-month-old fennec fox for sale, Tunisia, 2014

Arthur Tanner, girl with wolf for the *Daily Herald*, 1958

John Drysdale, *Descending a Staircase*, 1976

Joseph Maria Eder, X-ray experiment with frogs, Vienna, 1896

Professor Dotterweich, X-ray stereograph of a lobster, 1900

Sarah Anne Bright, catshark egg case, 1840

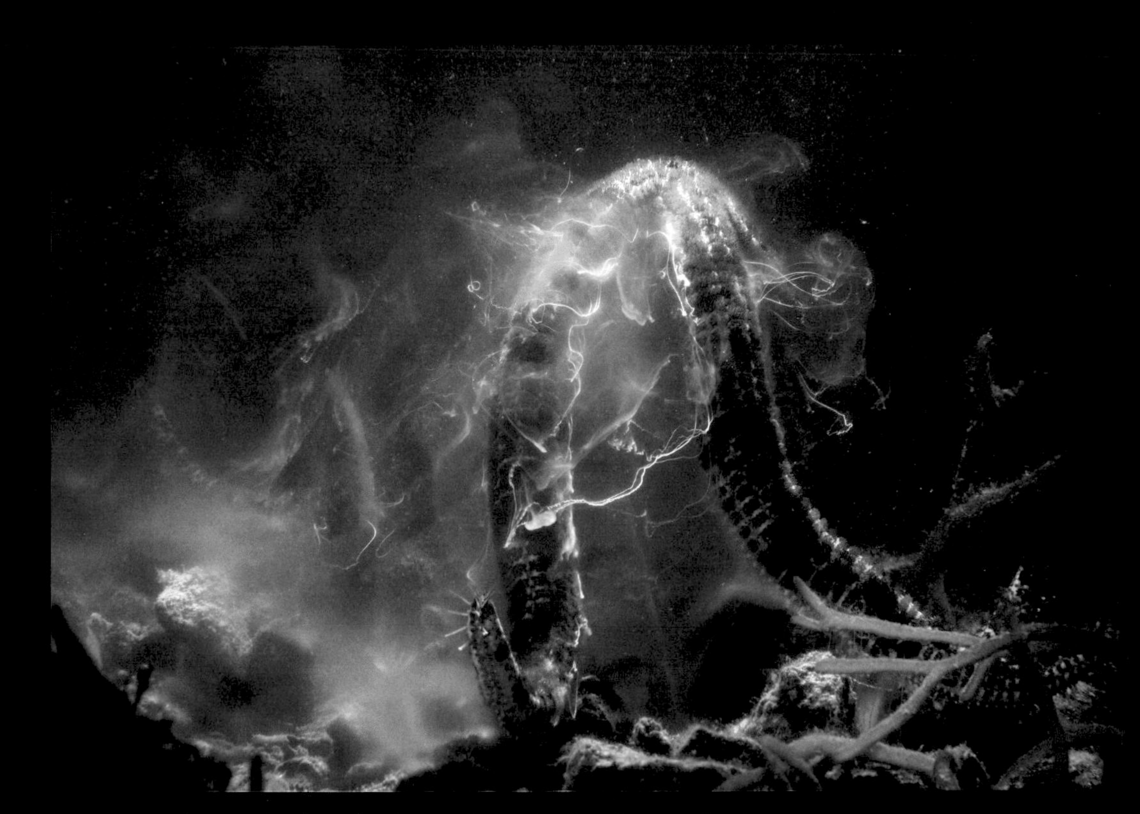

Tony Wu, Leach's sea star 'broadcast spawning', releasing streams of sperm
from its arms, Kagoshima Bay, Japan, 2021

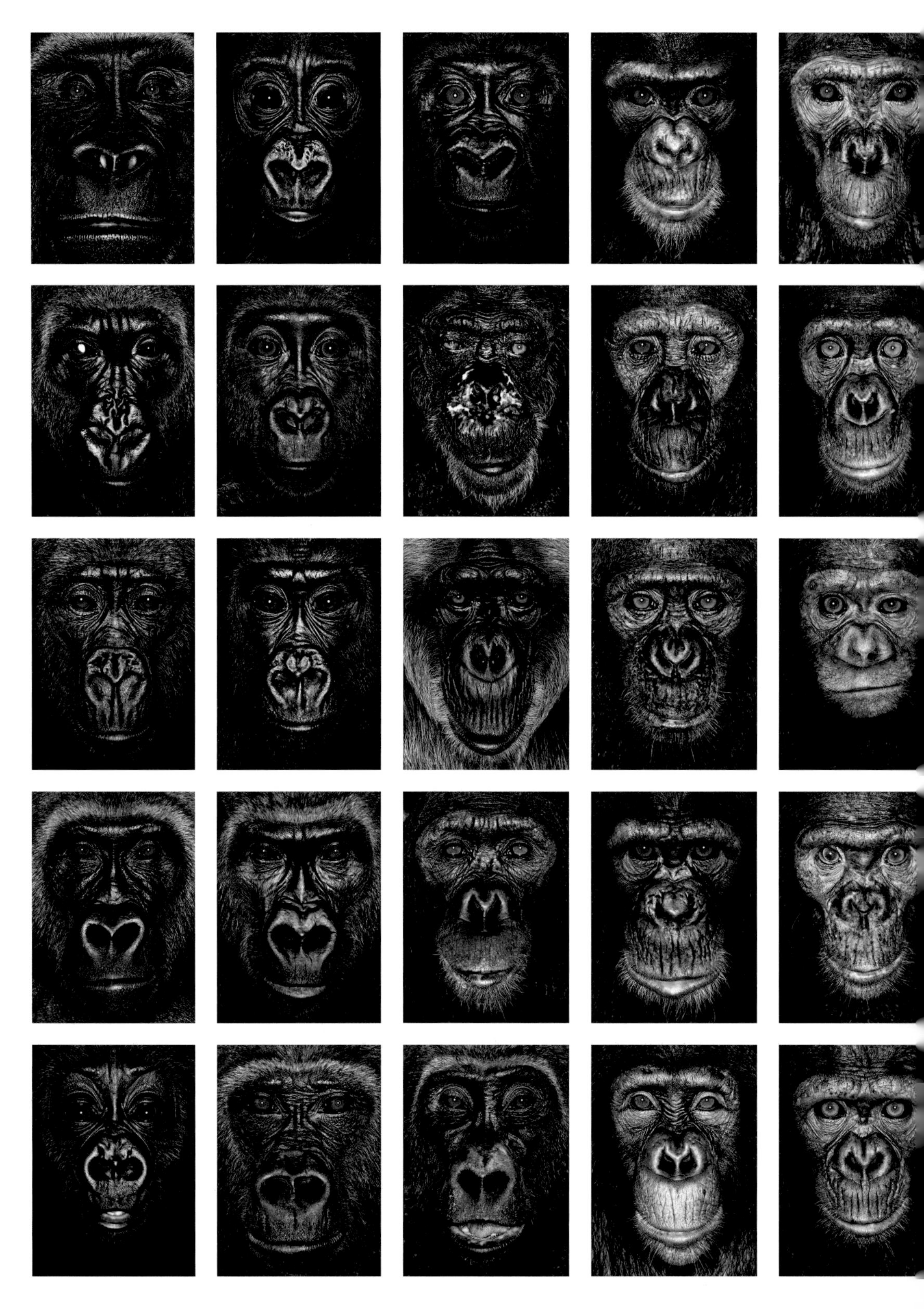

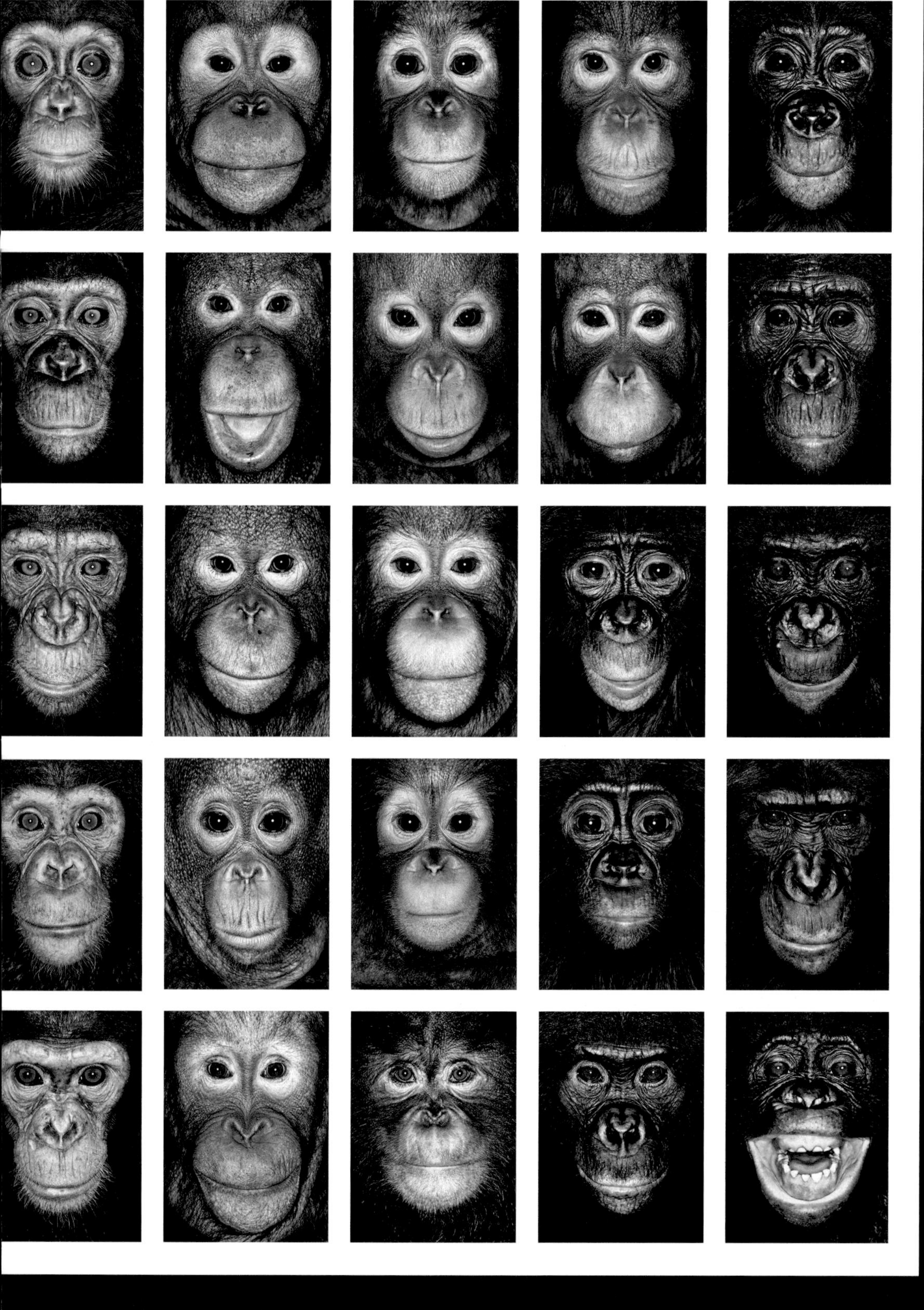

James Mollison, portraits of gorillas, chimps, orangutans and bonobos, *James and Other Apes*, 2004

HISTORIES

Wondrous Life

'In all things of nature there is something of the marvellous.'

Aristotle

ON 12 SEPTEMBER 1940, teenager Marcel Ravidat is walking his dog Robot in the woods outside the town of Montignac in south-western France, where for years there have been rumours of buried treasure. Scampering along the hillside chasing rabbits, Robot suddenly disappears from view.

The dog has found an unusual hole next to an uprooted tree. Marcel enlarges the opening and slides into a dark chamber beneath the ground. The treasure they uncover is a collection of ancient artwork dating back at least 17,000 years. Marcel would later say that he could see a 'cavalcade of animals larger than life painted on the walls and ceiling of the cave – each animal seemed to be moving'.

The Lascaux Cave, as it is now known, is home to some of the earliest examples of man-made art. There are horses, stags, ancient aurochs and bison, a bird, a bear, even a woolly rhinoceros. It could be said the discovery of these remarkable animals, these incomparable works of human visual culture, is all thanks to a dog.

A few years later, once peace has returned to war-torn Europe, a photographer named Ralph Morse heads to Montignac. His goal is to make the first colour record of the site, which experts are already praising as 'the Versailles of prehistoric man'. No easy thing so soon after the war. He has to dig a road bed near the cave to get all his equipment through the woods, installing a special generator shipped from London to provide light in the cave's narrow pitch-dark corridors. 'It was a challenging project,' Morse recalls, 'running wires in there, lowering all the camera equipment down on ropes. But once the lights were turned on ... wow!'

When his photos are released around the world in 1947, the reaction is staggering and the interest global. The caves open to the public a year later, but in 1963 they are closed when experts determine that carbon dioxide from the breath of thousands of visitors, as well as spores and other contaminants, are damaging the paintings. Today, only a handful of people are allowed inside Lascaux for a few days each year to monitor damage – 'a mysterious, encroaching mould' is the latest threat – while they work to keep the magnificent animals adorning the walls from vanishing entirely. It might be reasonable to ask whether the same care and attention is given to living animals.

The animals of Lascaux Cave are revealed to the world, press photograph, 1940
–
Seated on the right is Marcel Ravidat, who was the first to discover the cave thousands of years after these wondrous creatures were painted there. There are more than six hundred depictions of animals including horses, cows, bulls, stags, wolves, bears, lions, even a rhinoceros.

Ralph Morse, multiple exposure of a gibbon, 1965

–

Published as 'Private Life of Primates', a two-part photo essay in *LIFE*, February 1965.

Morse is working for *LIFE* magazine. It has a special place in the history of twentieth-century photography, so it deserves a closer look here. The magazine had actually been around since the 1880s, when it was a cheerful kind of general interest weekly, but as the Jazz Age turned into the Great Depression it lost money and subscribers and was sliding into financial ruin. It needed a revamp to meet the spirit of the times, and the man that takes on that challenge is publisher Henry Luce. He is convinced that pictures can tell a story instead of just illustrating text. Not just adornments: images *are* the story.

Relaunched in 1936 (and now styled in capitals), *LIFE* becomes the first all-photographic American news magazine and it dominates the market for decades, selling more than thirteen million copies a week in its heyday. Its influence on visual culture is profound, and its legacy for photojournalism, undeniable. The magazine runs as a weekly until 1972 and as a monthly until 2000, when it falls victim to the economy and finally folds. For decades its pages are home to some of the best photography in the industry. Morse is just one of hundreds who work for the magazine through the years, but his career is a great example of the sheer range of subjects and stories that photographers are tasked with recording.

He was as at home in the studio as in a cave, a ball pen or at a rocket launch. Morse just loved to create and was confident enough to improvise, like all the best in the game. Boxing kangaroos, *no problem*. Astronauts, *sure thing*. Hollywood film stars, *why not!* To shoot footage of rocket launches he used remote camera techniques of unprecedented complexity. To lure gibbons along a rope as he shot them for the first time with strobes, a simple bunch of bananas did the trick.

Animals are everywhere in the new pictorial magazines, whether in sport, business, politics or leisure. As symbols, in science, in hard news

Fritz Goro, queen triggerfish, 1953 and Nina Leen, green tree viper, 1963
–
The *LIFE* project 'Creatures of the Sea' was more than two years in the making, with Goro working above and below the water from Australia to the Bahamas. Leen's cover story 'Fearsome Fascinating World of Snakes' used photography to explore human feelings for these much maligned reptiles. In Florida, she shot an anaconda underwater using a special boat; closer to home, she hatched corn snakes in her bedroom.

or just for laughs, the animal image is a reliable visual commodity. While the purist photo historian might declare that the first image in *LIFE* magazine is a human baby, seconds after its umbilical cord is cut, wailing for breath – with the fitting caption 'Life Begins' – actually the *very* first photo inside the magazine is of an orangutan, a visual joke for an Ethyl motor fuel advert.

If you take time to look, animals are easy to find in that first edition: a trio of champion field dogs promote American whiskey; a family flees a tornado clutching their pets and circus elephants stand ready for their next performance in reproductions of work by artist John Steuart Curry. There are travel photos aplenty, a sleepy village in Brazil with a dog in the foreground that, the caption hastens to tell us, 'has fleas'. There are shots of Laysan albatross on Midway Atoll in the Pacific and the jolly air crew who've just whacked an impromptu golf course across their nesting sites. There are horses from the new Warner Bros motion picture *Charge of the Light Brigade*; and a black-and-white Dalmatian (in full colour) for a back-page ad for Lucky Strike cigarettes.

So, lots of animals in that very first issue of *LIFE*. But, to my mind, the highlight is in the heart of the magazine: a spread on spiders! Black widow spiders, to be specific, 'a perennial news story'. George Elwood Jenks has 'recently made this deadly insect his hobby' we are told, producing an impressive set of images. He photographs the animals that everyone loves to hate because he's curious to know more about them. Not for the spin that others would put on the images, but because of his own fascinations in the way they live, feed and reproduce. 'Can any brow-beaten husband', the editors ask, 'follow the sad career of the Black Widow's mate without a fellow feeling for the poor little creatures?' Cast the tone of the question aside, but think more on the idea: just how can photos help us think about animals with kinder eyes?

Through the decades, *LIFE* is filled with remarkable animal images. In the early days, naturally, there are lots of stories about dogs, farmyard animals, prizewinning pets. But as the magazine grows, so does its budget. Fritz Goro leads a team creating remarkable colour stories on coral reefs and sea turtles and goes on expedition to the Arctic; Nina Leen takes on poisonous snakes, then vampire bats; Stan Wayman treks through Indian jungles in search of tigers; John Dominis follows the big cats of Africa, then humpback whales; and Co Rentmeester heads off to Borneo to tell the story of threatened orangutans. With some squeaky toys and a pocketful of treats, Francis Miller drives to Washington to meet a pair of frisky beagles. He gets his shot and it makes the cover – President Johnson's puppies are a hit.

In 1958 *LIFE* begins a new series of features to commemorate the upcoming centenary of Darwin's theory of evolution. Photo teams are sent out to 'revisit the wild places and strange creatures that led to this momentous concept'. In 1960 the magazine publishes a book, *The Wonders of Life on Earth*, just in time for Christmas, and in little more than a year it sells over 230,000 copies. The *LIFE*

'Nature Library' series, six books by then, surpasses a million copies. This really is a photo industry on an unprecedented scale.

'Dogs in America' is one of countless articles about dogs, no surprise as the most popular animal of all, but this 1949 feature was new: entirely in colour. For much of its history *LIFE* is a largely black-and-white publication, particularly in its role as chronicler of the world's news – a role that put a premium on speed, and printing in colour halftones was still a slow and expensive process. A generation that had grown up with radio was used to hearing about news they couldn't see, and *LIFE* filled that void, but as televisions take off they have a problem. In the 1960s, at great expense, the editors decide to compete with television by using more colour and they have success, but by the 1970s the weekly becomes a monthly as budgets tighten. Despite challenges to the industry, their photographers are still creating work that informs and entertains, shocks and intrigues. They continue to take advantage of improvements in photographic technology to tell the story of the world around them with compassion, sensitivity, curiosity and courage.

Nina Leen is another good example of this combination of skills. She lives in Germany, Italy and Switzerland before moving to America, where she becomes one of *LIFE*'s first female photographers in the 1940s. Her debut images in the magazine are tortoises in the Bronx Zoo, but by the end of her remarkable career she has shot more than fifty cover stories. She also becomes obsessed with bats: 'I wanted to show these wonderful

Nina Leen, spear-nosed bats in flight, Kline Biology Laboratory, Yale University, 1968

–

Leen found bats frightening when she first encountered them, but when she started to study them more carefully she 'found them so engrossing' that she knew she must eventually make a book. She worked in zoos, science labs and a huge maternity cave in Texas where millions hung from the roof suckling their young.

'Knowing does not guarantee caring, but you cannot care about something if you do not know it exists.'

Sylvia Earle

creatures in pictures', she says, 'because words just fly off into the air.' She creates a bestselling book about bats that does much to rehabilitate their creepy reputation. She makes books on snakes and monkeys too and, predictably perhaps, one about cats and another on 'dogs of all sizes'. *And Then There Were None* documents America's vanishing wildlife in 1973 and 1977's *Taking Pictures* aims to encourage children to pick up photography. But perhaps the single assignment that has the most lasting effect on her life and work was actually one her colleague Leonard McCombe was shooting.

In 1949, he is covering a story in Texas when he comes across a puppy, flea-ridden and barely alive, resting on the body of its dead mother. Unwilling to simply abandon the creature, McCombe ships it off to the *LIFE* offices in New York, where Leen (who was known for liking animals more than she liked humans) adopts it. In no time at all the dog, dubbed 'Lucky', becomes America's pet. Nina brings Lucky with her everywhere, documenting the dog's post-rescue adventures in follow-up articles, a short film and her most widely read book: *Lucky, the Famous Foundling* (1951). It is 'one of the greatest human-interest stories of the century', so runs the publisher's spiel, which just goes to show the appeal of dogs and the power of a good story.

Award ceremony for the Wildlife Photographer of the Year competition, 1966

–

David Attenborough (left) congratulates Roger Dowdeswell, whose colour photo of a tawny owl won the inaugural competition.

* * *

'I'm not an animal lover if that means you think things are nice if you can pat them, but I am intoxicated by animals.' The words of David Attenborough, who needs little introduction, arguably the greatest wildlife broadcaster of all time. So many of the projects he has been part of set the benchmark for quality and innovation and influenced hundreds, if not thousands, of wildlife film-makers to progress in the industry. His enthusiasm, knowledge and, most of all, his respect for nature has won him millions of admirers around the world.

In the 1950s, when Attenborough begins his career, much of the world's wildlife is still unrecorded, even unknown. It is a golden age of nature film-making in many senses, with teams finding new ways each year to show animals in their natural habitats. Austrian biologist Hans Hass, a diving pioneer, and his wife Lotte are bringing many of the wonders of the under-sea world into homes for the first time, and Disney are putting animal stories on the big screen, with many of their documentaries winning Academy Awards. And yet, the origins of life and the structure of DNA are still a mystery; the idea that continents could drift across the surface of the planet, ridiculed. Science was something mostly done in laboratories or museums, but with war now over, a new generation of researchers head out into the field. Attenborough's vision is to translate their insights into imagery and films that can combine education and amazement.

It is estimated that some five hundred million people worldwide watch his thirteen-part series *Life on Earth* – launched in 1979 – which he writes and presents, and which is by far the most ambitious show ever produced. His team film and photograph 650 species and travel to 39 countries, and there it all is in colour! A sequel, *The Living Planet*, comes five years later in 1984, and his study of animal behaviour, *The Trials of Life*, reaches screens in 1990. The remarkable footage of killer whales hunting sea lions on the beaches of Patagonia and chimpanzees fighting colobus monkeys draws strong reactions. More specialized surveys follow, hand in hand with developments in camera technology and logistical possibility.

Life in the Freezer in 1993 is the first television series to devote itself to the natural history of Antarctica; *The Private Life of Plants* two years later showcases time-lapse photography; *The Life of Mammals* in 2002 is equally pioneering in its use of low light and infrared, revealing the unseen behaviour of nocturnal mammals; and in 2005, in *Life in the Undergrowth*, advances in macro-photography make it possible to capture the world of invertebrates for the first time in exquisite detail, from Australian redback spiders to venomous centipedes in the Amazon.

He is humble enough to admit that his teams of camera operators take on the lion's share of effort, often in implausibly difficult conditions, or waiting patiently for weeks, even months, to get the shots they need. Yet sometimes, just one person with a good camera and the right natural history knowledge is all it takes. He points to the work of Doug Allan, persevering alone in a hide, high in the Himalayas, to get the first ever video footage of snow leopards for *Planet Earth*.

Attenborough's career has been a long one, widely acclaimed and, thanks to superlative teams, always pushing the possibilities of the medium. His achievements span broadcasting media – he is the only person to have won BAFTAs for programmes made in black and white, colour, HD and 3D. His camera operators have devised special filming techniques to obtain footage of elusive animals, from rare frogs to bats in flight, to that memorable encounter with mountain gorillas in Rwanda in 1978, which he still regards as one of the special moments of his life. Since then, newly discovered species have also been named in his honour: an Ecuadorian flowering tree, alpine hawkweed, a flightless weevil, a spiny anteater, a Madagascan ghost shrimp and *Attenborosaurus*, a prehistoric plesiosaur.

So, why take photos or make films about animals? 'People must feel that the natural world is important and valuable and beautiful and wonderful and an amazement and a pleasure', he says. 'It seems to me that the natural world is the greatest source of excitement; the greatest source of visual beauty; the greatest source of intellectual interest. It is the greatest source of so much in life that makes life worth living.'

(below) Koko, a female western lowland gorilla, self-portrait on the cover of *National Geographic*, 1978

–

Born at San Francisco Zoo, Koko was taught sign language from an early age by researcher Francine 'Penny' Patterson. Koko learnt more than a thousand signs and could understand more than two thousand words of spoken English. It was said she was paid $750 by the magazine for the use of her selfie. Apparently, that bought her a lot of fruit and – her favourite – nectarine yogurt.

(above) Ami Vitale, *Sudan*, 2018
–
Joseph Wachira, a wildlife ranger, comforts the last living male northern white rhino, moments before he passed away at Ol Pejeta Wildlife Conservancy in Kenya.

Sudan, the last male northern white rhino on the planet, is comforted by Joseph Wachira moments before he passes away in March 2018. Photographer Ami Vitale captures a touching image of the scene, commissioned by a magazine with readers in every country on earth. 'It felt like watching our own demise,' Vitale explains. 'This giant creature had survived as a species for millions of years but could not survive us, mankind.'

The magazine is *National Geographic*, which has inspired and supported so many of the photographers in this book. It is second only to Attenborough perhaps, as a major early stimulus. Growing in pre-eminence, even as *LIFE* folded, it carries some of the best animal photos in the business. A feature in the magazine at the end of 2020 raised interesting questions: what were their most compelling images of the twenty-first century so far? And how best to tell the stories of the century in pictures? Narrowing the photos down must have been a tough edit – by their own estimates, the organization made more than 1.7 million pictures that year alone.

Interestingly enough, many of the images selected for this best-of-the-century round-up are of animals. No surprise with *National Geographic's* audience in mind, but considering photography's evolution from daguerreotype to digital, it is amazing that animals are front and centre. Leading the photo feature online is a shot of a curious 3-metre (12-foot) leopard seal hunting young penguins in the frigid waters off Antarctica, something almost inconceivable to capture in images until recently. The seal engulfs Paul Nicklen's camera – and most of his head – in its gaping jaws. 'That image and story changed my life,' Nicklen says. 'It allowed me to use love, humour and adventure to get people to dismiss preconceived ideas about misunderstood predators.'

Photographs of animals enable us to tell new stories and change the narrative of things we take for granted. They capture our attention and demand that we take more notice. But how long is it before we look away again? In 2007 Brent Stirton's photographs of slain gorillas in Congo's Virunga National Park – casualties of the region's illegal charcoal trade – ignite global outrage. Today these gorillas remain endangered by climate change and continued loss of habitat. Perhaps more encouraging is Charlie Hamilton James's shot of a grizzly bear feeding on a bison carcass in Grand Teton National Park. It is truthful and inspiring: not cute and cuddly, but beautiful nature. Though bear numbers fell to as low as six hundred in the 1960s, they had climbed again to nearly one thousand by 2010 due to protections afforded by the US Endangered Species Act.

Anand Varma's image of a male sheep crab demands a closer look. This isn't an ordinary crab but rather a zombie crustacean that has been invaded by a parasitic barnacle. Varma has spent years capturing the world of mind-controlling parasites like this, which uses its powers to widen the crab's abdomen, creating a womb for the parasite to fill with its own eggs. Photos enable wondrous new visions, rich in the realities of nature. They allow us to go places and witness things we could barely dream of seeing.

Perhaps the most famous of all the animal photographs this century, is Vitale's portrait of Wachira and the rhino Sudan. Vitale first met Sudan in 2009 and has since dedicated herself to documenting the plight of the subspecies – driven nearly to extinction by poachers who treasure its horn. The power of the *National Geographic* platform extends way beyond the print magazine or exhibitions of old, of course. The @NatGeo Instagram account was the first brand – other than Instagram itself – to pass the 25 million-follower mark in 2015 and it now has over 200 million followers. A single post of Vitale's rhino on its feed reaches an audience that would have been unimaginable to nineteenth-century photographers.

That post created much-needed attention not only to the crisis of extinction but also to the important work being done by people like Wachira, the wildlife ranger. Photographs narrate the story of this loss and take time to celebrate small successes too. Today, there is a glimmer of hope for the northern white rhino. Two females remain, and scientists are boldly attempting to revive the rhino population via in vitro fertilization; using technology to try to right the wrongs of centuries of destruction.

(opposite) Brent Stirton, rangers and locals carry the body of a mountain gorilla, Virunga National Park, 2007

–

A total of seven gorillas had been slaughtered by people involved in the illegal charcoal trade. They were after the valuable hardwood trees in Virunga's protected areas and killed the gorillas purely to try to intimidate the rangers.

(overleaf) Charlie Hamilton James, a male grizzly chases ravens from a bison carcass, Grand Teton National Park, Wyoming, 2015

–

The grizzly populations of the Grand Teton and Yellowstone National Parks have been protected since the 1970s, but now that numbers are recovering it has been proposed that some hunting be allowed outside the National Parks. This raises concerns not only about the bears' fate but also the wider ecological consequences.

Xavi Bou

XAVI BOU *is known for his project* Ornithographies, *which makes visible the beauty of bird flight paths. His work has been published in magazines like* National Geographic *and major European newspapers. He has exhibited* Ornithographies *in Australia, the Netherlands, the USA, Spain, Switzerland, France, Russia and Greece.*

Do you have a favourite animal photo that you have taken? Probably my pictures of starlings: the complexity and the beauty of flocks in flight. The interaction with the hawks and the relation between each individual is so fascinating to me.

Is there one historic shot that you really admire? I think of the George Shiras night pictures [for *National Geographic*] – for me, they represent the magic and the exploration of the unknown. Going to the edge of the technical possibilities they had at that time, he got these magical images.

It is often said that 'photography can change the world'. Do you agree? Photography can change the world if it achieves change in the consciousness and habits of the human animals who are in control here. It is so often said, but it's worth repeating: you can't protect something you do not love, you can't love something you do not know.

Is there an animal photo that first changed your world? Thanks to pictures of the richness of the forest, when I saw fires destroying it to clear land for cows I changed my eating habits. That's a pretty immediate influence for a photo, isn't it! I try to eat less meat now and make sure when I do that it comes from well-managed farms local to me.

Are animal photos important? Photography is a tool that people have been using for more than a hundred years to show, in general, the things that are important to them, be they positive or negative. The rise of the photograph mirrors the continuing interest of the animal world in society and yet we are treating animals worse than ever. Many people are getting closer to nature thanks to photography. Some of them until a few years ago would have chosen a rifle instead of a camera – and this is good news. But it is not the end of animal cruelties.

Do you have any other thoughts to share? The big difference between chronophotography and my work: the first is a study of movement where you can see the position of the body of the animal in each moment, so you could examine it. In my case, I shoot more pictures per second so, between one image and the other, the images overlap and the silhouette of the body gets lost. At that point, it is very hard to know what you're looking at. So in concept, it shows movement in a clear way, in this case, *flight* not a bird flying. To me that's an important distinction. An image that is beautiful, beguiling and that is a trace of an animal. An image that is made by the animal. That is the animal but, at the same moment, is also *not* the animal. Something different, something new. I photograph animals to find new beauties hidden in plain sight.

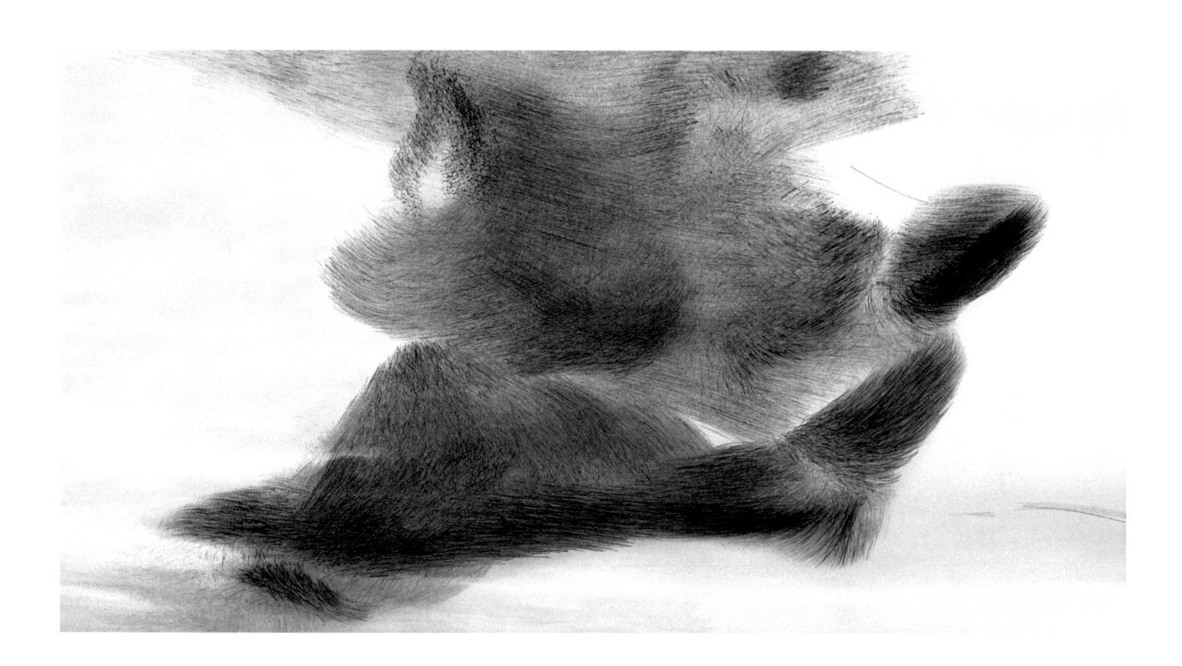

Common starling, Ebro Delta, Catalonia, 2019

(top) Common swift, Palafrugell, Catalonia, 2020
(above) Grey heron (with red kite on the tree), Lleida, Catalonia, 2016

‘I wish to show the hidden beauty of bird flight. In changing the perception of time, in revealing these patterns, I hope these images help to connect more people with the infinite beauty of nature.’

Xavi Bou

(above) Northern fulmars, Skógafoss waterfall, Iceland, 2017
(overleaf) Tree swallow, Grand Teton National Park, 2019

Alexander Semenov

ALEXANDER SEMENOV *is a Russian marine biologist, specializing in invertebrate animals. He is head of the divers' team at Moscow State University's White Sea Biological Station, where he manages all sorts of underwater work. His key specialism is scientific macrophotography in natural environments.*

Describe an animal photo you've taken that was memorable to you. I have a picture of a giant octopus, taken five seconds after my buddy ripped it off himself. The octopus had hugged him completely, so only the fins and scuba tank were sticking out. Before that, the octopus attacked me by grabbing my leg tightly and quickly beginning to climb towards my head, but I kicked it with my free leg and freed myself. While I was changing the camera settings and circling around to get a shot, my friend was also wrapped up by it.

Describe an animal photo that you've missed or not taken: Mating sea angels. We dived under the ice for three months, shooting a story about pteropod molluscs for a TV documentary. We had shot a lot of footage already, but never encountered a mating.... And so, in my beloved White Sea, after three months of work, I dived without a camera to collect live material for a scientific request and met as many as four pairs of mating angels!

What is the most influential animal photo ever taken? That's an impossible question. But we could think about it from different directions. David Slater's 'Monkey selfie', for example, has led to really wide-reaching discussions about copyright and ethics. It's also a funny image and it has spread all over the Internet. The crested black macaques were much ignored, but now tourism may help to protect their habitat. Far more meaningful, Brent Stirton's *Memorial to a Species*, the dead rhino. Elephants in chains. Polar bears on tiny pieces of sea ice. They can each affect us, and change things, in different ways – or not. Many great images just get lost in the noise of life, with other things shouting more loudly for attention.

So, why do you photograph animals? I photograph animals to study the world we live in. To show what I see to other people who are also curious and interested. To tell stories. To impress and inspire young kids to become scientists. To fill in the gaps in our knowledge. We still have so many of them that there is enough work for several generations to come. To make science truly beautiful. To make my life truly beautiful. I could give a thousand and one reasons why I like doing this. I love it, and I can find as many meanings as one needs.

Can you describe the positive impact that one of your photos has had? Ah, that's easy. There are several people who enrolled in the biology department at my university because they saw my photos and read all the stories I write online – and in the end they wanted to study the underwater world!

It is often said that 'photography can change the world'. Do you agree? Well, no. I think only knowledge can change the world. Photography can show this world, can show what to change. It can provoke curiosity, a desire to know more. But without understanding what needs to be changed, the world will not change for the better.... In our overloaded media world, I don't rely much on the power of photography. I look through and immediately forget hundreds of photos every day. On the other hand, for example, my jellyfish photo turned out to be a door for someone into marine biology. And certainly that person's world has changed. You can look at it from different angles.

Lion's mane jellyfish in the strait of Velikaya Salma, White Sea, 2019

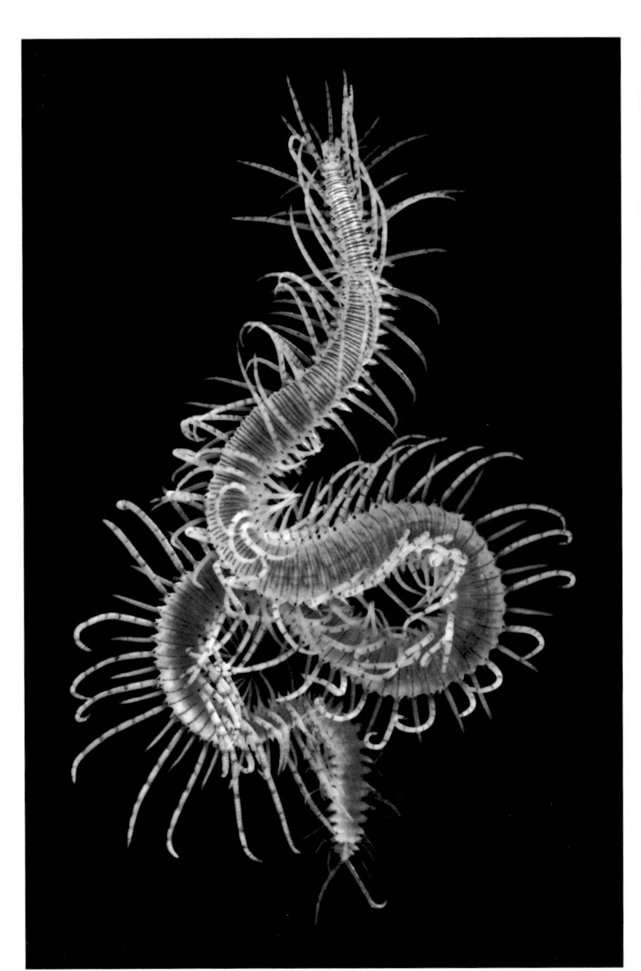

‘Animal photography documents the diversity and beauty and weirdness of the world we live in. Many people do not have the opportunity to see with their own eyes what photographers spend their lives on. Photographs provide that chance.’

Alexander Semenov

(above) Syllis maganda and *Lanice viridis,* two species of polychaete marine worm, Lizard Island Research Station near the Great Barrier Reef, Australia, 2013
(opposite) Sea angel, White Sea Biological Station, Russia, 2016
(overleaf) Giant Pacific octopus, Vityaz Bay, Sea of Japan, 2012

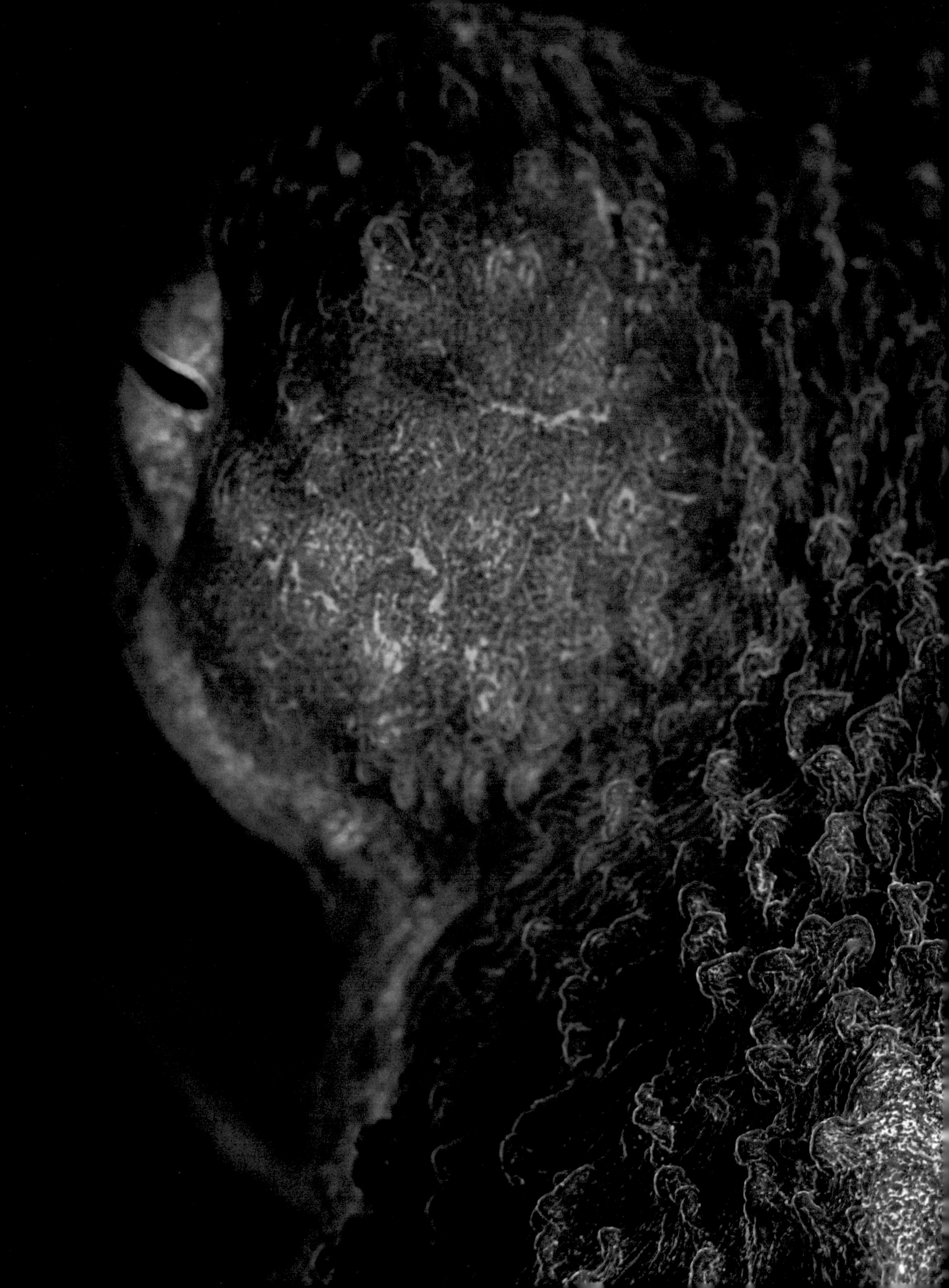

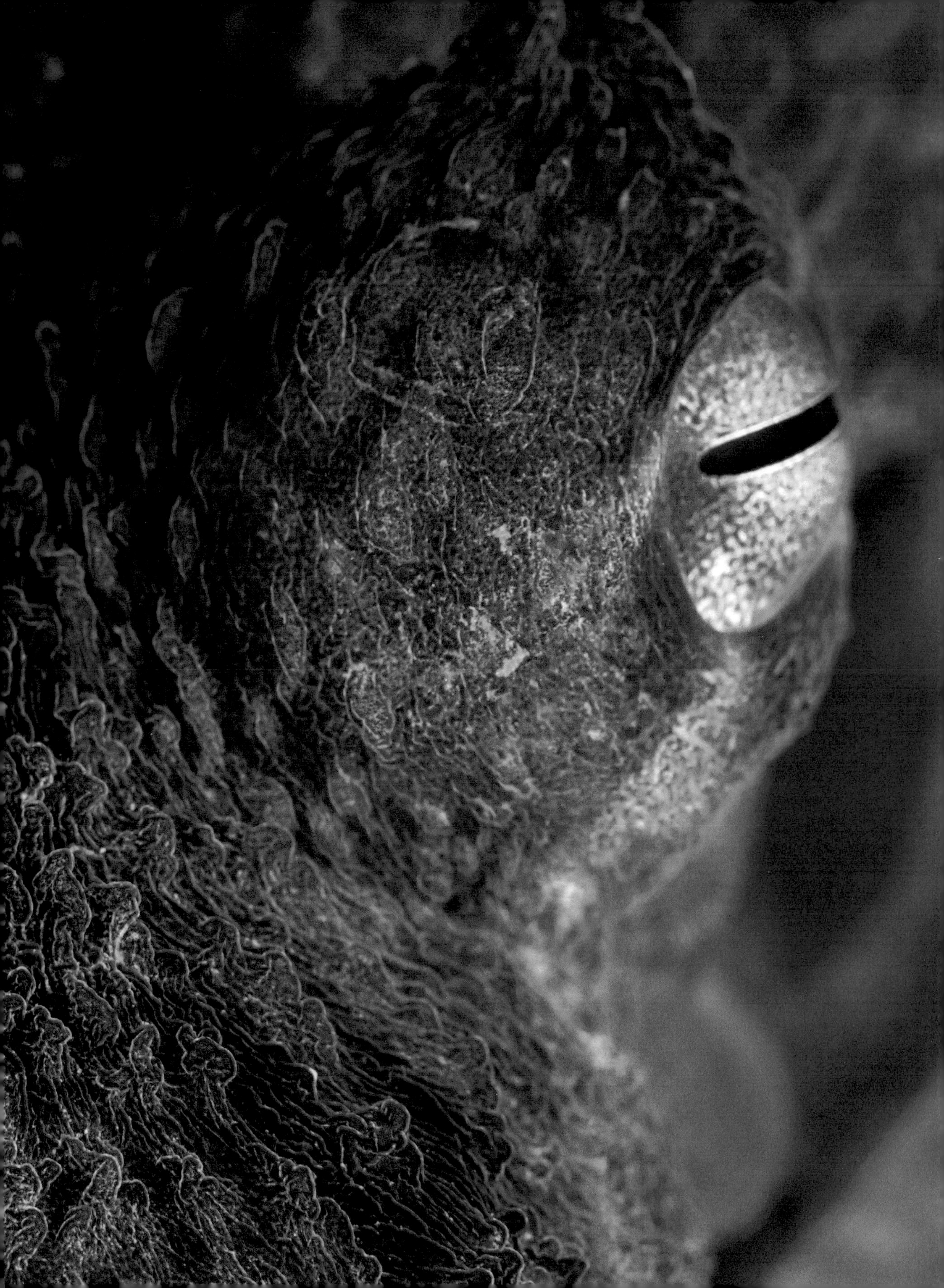

Sergey Gorshkov

SERGEY GORSHKOV *grew up in a Siberian village. In his thirties he discovered photography as a way to re-engage with nature. He sold his business and took up photography, determined to record the richness of Russia's wilderness. In 2020 he was awarded the Wildlife Photographer of the Year Grand Title.*

Is there a particular photo, or group of images, that inspired you into action? Nature is the daily inspiration for me, not the work of others. I am always looking for something new. Each time I go into nature I try to see new things and you must try to always be inventive. But sometimes, just being out is enough. Why do I love it so much? The simplest explanation is the opportunity to listen to the roar of a leopard in the delta of the Okavango River, or the cackle of flying geese across the expanses of the Taimyr tundra, or standing on a cliff on Wrangel Island and feeling the bite of an Arctic wind, or the heat of hot lava on the volcanoes of Kamchatka.

Have you ever risked your camera for a photo? Your cameras have to withstand the most difficult situations and weather conditions. And sometimes be gnawed and clawed by lions and bears, or be stolen at the airport.

Have you ever risked your life for a photo? I suppose I have many times. If a bear wants to eat you it will. They mostly don't, but you can never be sure. Every bear is different. You have to try and avoid this scenario.

What do you hope to achieve? I try to shoot where no one has shot before. I try to shoot at times of the day when few have got good images. I want to find things that I have never witnessed.

Has an animal photo forced you to examine your life? The moment I went to the Arctic for the first time, there were enough real images from nature for me to see. I realized immediately that I must give all my energy to creating images of the animals that survive in this wonderful region.

Can you describe the positive impact that one of your photos has had? When I published my first books on bears, I began to receive letters from hunters. Many of them stopped hunting, changed their carbine to a camera and started taking pictures of wildlife. I'm glad that my photos could change their world for the better too. Many years ago on Wrangel Island, I filmed a polar bear walking near old, rusty gasoline barrels. I published this photo in a magazine and made a great post on my social networks. Fortunately, this photo and the problem were seen by the Russian government and it helped to start the process of clearing the Russian Arctic of a huge number of empty barrels.

It is often said that 'photography can change the world'. Do you agree? I have become a witness to the negative effects of human intervention on wildlife. My anxiety grows every year. The world that I am filming is in danger. Abundance diminishes. Many wild animals and birds have disappeared, many are on the verge of extinction, many will disappear whether we like it or not. But I am also witness to remarkable human efforts to protect species in their habitats. In our world, photography now is a powerful tool in nature conservation. Photos affect the consciousness of people and they begin to look at nature in a different way.

(above) Arctic fox stealing a snow goose egg, Wrangel Island, 2011
(overleaf) A polar bear walks along the glacier front, Russian Arctic National Park, Franz Josef Land, 2017

Amur tiger leaving its scent for potential mates, Siberia, 2019

PROFILE

Tim Laman

TIM LAMAN is an American ornithologist, wildlife photojournalist and film-maker. He has documented all the species of bird-of-paradise in their native habitats during research expeditions with the Cornell Lab of Ornithology; work that was first publicized in a 2007 National Geographic *article.*

Can you remember the first photo of an animal that you took? Black kites when I was a senior in high school in Kobe, Japan. I climbed a tree to shoot a chick in the nest. I still have the photo.

Is there a particular photo, or body of work, that raised your curiosity like no other? David Doubilet's work raised my curiosity about the ocean world for sure. Mark Moffett's amazing macro work on ants and other critters opened up a whole new world.

So, why do you photograph animals? I love exploring wild places. Photographing animals gets me out there with a mission. I also love the 'thrill of the hunt' and the creative artistic challenge: trying to make an image no one has seen before, capturing a unique behaviour in a spectacular way or revealing the unexpected.

My photo of a greater bird-of-paradise at sunrise in Indonesia back in 2010 is the best example. There have been rare moments in a long career in the field, when the image I had pre-visualized in my head came to life and in fact exceeded my expectations. This shot is one of those moments. Every day I climbed, in the pitch black before dawn, first the tree where the bird had been coming to display to rig a remote-controlled camera camouflaged with a wrapping of leaves; then I descended and climbed a second, neighbouring tree while it was still dark and perched in a little blind with my laptop, connected by cable across the canopy to the camera. When the male arrived that morning it was before sunrise. But he stayed, calling and displaying and, when the sun popped out and lit up the bird and the mist in the canopy, I saw the scene come to life on my screen and I made the image. When you are in the field and in the heat of the moment, you often aren't sure exactly what you have. But this time I really knew at that moment I had captured a very special image. I can still remember how my heart raced. This image has been a double page in *National Geographic* and won top honours in the Wildlife Photographer of the Year competition, but I'm most proud of the role it has played in helping to elevate birds-of-paradise as the ambassadors for the conservation of rainforests in New Guinea. It took me almost ten years to make this shot.

Are animal photos important? I know it's a bit of a cliché, but people only appreciate what they know about. If no one has ever seen images of New Guinea's birds-of-paradise, they are going to care less about those forests being destroyed. As for random animal photos vs award-winning wildlife photos or wildlife photojournalism, my experience from social media is that people on the Internet actually appreciate quality photography. Award-winning photos get a lot more engagement than run-of-the-mill ones. So I think quality photography is still hugely important to get people's attention. Henry David Thoreau wrote, 'it's not what you look at but what you see.' He wasn't talking about photography, but that's how I think about photography. You look at everything out there in the world, but what do you decide to put a frame around and turn into a photograph? It's a strangely satisfying thing to do.

What does photography mean to you? Photography is a means to explore, document and reveal the unknown wonders of the natural world to people who will never experience them first hand.

A Bornean orangutan climbs a fruiting strangler fig,
Gunung Palung National Park, Indonesia, 2015

(opposite, above) Victoria's riflebird, Atherton Tableland, Australia, 2008
(opposite, below) Western parotia, Arfak Mountains, West Papua, Indonesia, 2009
(above) Lesser bird-of-paradise, Oransbari, West Papua, Indonesia, 2009

Melissa Groo

MELISSA GROO *is a photographer and writer, committed to telling the stories of wild lives in service to their conservation. She is passionate about wildlife photography that combines artfulness and authentic natural history, to evoke wonder and empathy.*

Is there a particular photo, or group of images, that inspired you into action? Any taken by the photographers who are members of the International League of Conservation Photographers. Any photo that simultaneously shows an animal and the threat to it. It's a real art, to create a frame that combines both well, and one I'm still learning how to achieve. I don't imagine I will ever be done learning how to create such images.

What do you hope to achieve? In terms of the field of wildlife photography, I want wildlife photographers to think very carefully about their impacts when they're in the field. Ethics in wildlife photography is a huge passion for me, and I've worked with numerous organizations, publications and photo contests on the issue. I've written about the topic a lot, but I'm not perfect. I'm always holding my own feet to the fire. I make mistakes. But I think if we bring knowledge and empathy to bear on our decisions in the field, we're doing the best we can. And we better damn well use our photos to advocate for the animals that we just about always disturb when we're out there.

Can you describe the positive impact that one of your photos has had? I took a photo that has been used numerous times to educate people about the terrible toll that buildings take on migrating birds. It shows two scarlet tanagers. They're dead, and it's sad to look at, but also beautiful. It seems to inspire people to make change. I've heard many testimonies from people to this effect and the National Audubon Society and the American Bird Conservancy licensed the image to educate the public.

It is often said that 'photography can change the world'. Do you agree? Absolutely. I think photography is more powerful than it's ever been. A single image has the ability to go viral – to be seen within seconds through social media all over the world. It's astounding. And people don't really read much anymore. They look at images. Everyone's short on time and we are such visual creatures and we carry these little 'picture books' (our phones) with us everywhere. I think it's an exciting time to be a photographer. That being said, we are bombarded with images every day, and it's easy to get lost in the shuffle. You have to stand out and I think the best way to do that is to be authentic and to tell a story. It's also the case that nature photography is becoming one of the only ways people either see or learn anything about the natural world, due to the ever-increasing number of people living in cities. That's why it becomes ever more important for us as nature photographers to tell the truth. To tell conservation stories. To not deceive the viewer. To connect people to the real, natural world.

Are animal photos important? Photos of wild animals can effectively and elegantly marry art to science. This is hugely important because scientists for the most part have no clue how to reach the general public with their findings. They bemoan this all the time. Photography can be an incredible vehicle for this. I think that's really exciting and supremely important.

What does photography mean to you?
Photography is my love for animals made visible.

Bobcat mother and kit, upstate New York, 2015

‘I photograph to find peace and solace in nature. It’s the only time in my life when everything else falls away. It’s akin to meditation.’

Melissa Groo

(opposite) A least tern shelters two chicks beneath her wings, near Ocean City, New Jersey, 2018
(above) A pair of scarlet tanagers lie dead on a bed of clover, having flown into a glass door, New York, 2010

Dina Litovsky

City Pupparazzi

Dina Litovsky *is a Ukrainian-born photographer living in New York City. She describes her imagery as visual sociology: exploring the idea of leisure, subcultures and social dynamics. In 2020 she won the Nannen Prize, Germany's foremost award for documentary photography.*

Known for his witty street photos of dogs, Elliott Erwitt once said that he liked dogs because they don't mind being photographed and because they don't ask for prints. This might still be true of dogs themselves, but their owners, especially those who oversee their pooches' Instagram accounts, are a different story. When I was photographing dog parks for *TIME* magazine in 2017, I got called 'pupparazzi'. The times have definitely changed since Erwitt's anonymous canine photos; every self-respecting modern dog expects to be tagged on social media.

Before I continue, a confession: I'm a devoted cat person. Living with a canine in a New York apartment never seemed like a good idea. Yet, the East Village is dog central and every day I watch people picking up after their precious four-legged friends with a mix of admiration and bemusement. As I went out to shoot in dog parks, I started thinking about what would make a portrait of a dog interesting. Can we even consider an animal photo a portrait? I imagine it depends on which animal and the extent to which we can anthropomorphize a species. We tend to view cute animals as having greater complexity and empathize with them by assigning them human-like qualities. And we are more in tune with those animals we treat as pets, becoming familiar with expressions that we would otherwise miss on another species. It's probably easier to conceive of a cat portrait than one of a snake or a hippo.

Dogs are easy to anthropomorphize and we love doing so with them probably more than any other animal. William Wegman takes the best-in-show prize; his photography has exhausted every possibility of how to dress up his Weimaraners over a period of fifty years. The humorous images occupy a netherworld somewhere between a portrait, a fashion editorial and a meme. They could seem gimmicky at first, but under the layer of slapstick lies a tongue-in-cheek metaphor of the human condition. We are not laughing at the dogs but at ourselves reflected back to us in canine form.

After documenting three Westminster dog shows, I learned about more canine breeds than I thought possible. With each breed, and more importantly with each breed owner, came a slightly different approach. The majestic Afghan hounds are sensitive animals who don't love getting flashed in the face. It agitates them and their owners even more so. Apparently, the hounds are much less sensitive to noise. Pugs are friendly and inquisitive, but also kind of intimidating. I couldn't escape the feeling that they were doing me a favour by posing. They are also the breed most likely to have an Instagram account.

English bulldogs are resigned and chilled. Photographing them was easy since they just let me do my thing without even a slight acknowledgment of my existence. Vizslas are regal. I never knew this breed existed until I met Nova in Washington Square dog park; I felt like I was in the presence of the queen of England. Pomeranians are highly strung. Pit bulls are just

Pug and rooftops, New York, 2017

'I'm unable to take an interesting portrait of a cat because my desire to hug them results in nauseatingly sweet postcards; dogs are a different story. I'm curious to know more about them and I keep my cool.'

freaking awesome – smart and receptive, calm and curious. I found myself wanting to hug Tico, a two-year-old pit bull who patiently posed for me in a bed of tulips.

The infamous street photographer Garry Winogrand said that we should beware of pictures of dogs and children because they are not as good as you think they are. That perfectly echoes my cardinal fear when photographing animals, the inevitable pitfall of making cute photos. As it turns out, you don't have to be a dog person to make images of dogs. In fact, *not* being a dog person can probably save you from the dark force of adorable imagery.

* * *

When I was starting out in the photo industry, I was known mostly for my nightlife work. I became the go-to party photographer for several New York publications. For a couple of years, it was fun and rewarding, covering everything from rowdy neighbourhood bars to society balls. But suddenly I was entering my mid-thirties, the dark circles under my eyes from the constant all-nighters were getting fiercer, yet I was still stuck on the party bandwagon. Photography is a tricky business; magazines love to pigeonhole photographers. So, at first, you devote a few years trying to get noticed by being consistently and doggedly good at a specific genre. Then, after your reputation as a '___ photographer' is established, you spend the next decade attempting to unravel the singular focus and prove you can do other things as well. After five years of mostly photographing social events, I was clamouring for something, anything else ... even if it was shooting rats!

When I first got the assignment from the *New York Times Magazine* in 2018 I was a bit dumbfounded. I've never been particularly afraid of rats (after all, they are just squirrels with less stellar tails) but I couldn't imagine photographing one either. The project was a part of their annual 'Voyages' issue, that year focusing on the sounds of cities. Unsurprisingly, I got New York. But somewhat surprisingly, the noise chosen to represent NYC was almost entirely inaudible: the sound of rats talking.

When rats communicate, we can't hear much. Their range is at a different frequency, so a recording needs to be made using an ultrasonic microphone. Once the pitch is lowered to adjust for human hearing, a whole universe of sounds is revealed. Not only do rats constantly talk while socializing, courting or warning each other about danger, they also laugh. The laughter, coming in pulsating sound bursts, doesn't exactly sound human yet, incredibly, it is instantly recognizable as laughter. I also learned that rats love being tickled, which makes them giggle with abandon.

In the last few years, I've been able to get away from photographing parties and now get sent on all kinds of assignments, from big-wave surfing to presidential inaugurations, but the rat shoot remains one of my favourites. I gained a lot of respect for our whiskered little neighbours who, like us, are just trying to survive through the relentless madness that is this big city.

* * *

Another day, another glimpse into the wonder of animals. It began with a question, two in fact: If you bless an animal, does that mean it has a soul? And if it has a soul, does that mean you

Tico the pit bull in tulips, New York, 2017

Rat, for the *New York Times Magazine*, 2018

shouldn't eat it? These were the ideas running through my mind as I photographed the annual 'Blessing of the Animals' in New York's Cathedral of St John the Divine. It's not a new thing: they've been doing it for over thirty years, yet most native New Yorkers are unaware of it. Every October, animals big and small are brought to the Upper West Side to be blessed in honour of St Francis, their proclaimed patron saint. That year the procession included, among other creatures, a spider, a rat, a hawk, a cow, a horse, a couple of owls and a camel. A tortoise was wheeled in on a cart so as not to delay the ceremony with its unhurried pace. Afterwards, the priests gathered in the garden to bless pets brought by the attendees – no surprise, mostly dogs, some cats and a parrot. All in all, it was a very strange afternoon.

A disclaimer: I'm neither religious nor vegetarian. When it comes to food, I'm an 'occasiotarian'. In the past I have been a voracious meat eater, then, after reading a couple of books about factory farming, I became briefly but enthusiastically vegan, which was followed by being vegetarian (can't live without cheese and butter) and then compromising down to pescatarian (sushi would my last meal on earth). Finally, I settled on a sustainable compromise – eating a little bit of everything, while limiting meat to rare occasions.

Photographing the blessing ceremony reignited my conflicted relationship with the issue. The care put into handling every single creature in attendance was moving, but I couldn't escape the feeling that the animals were props in an elaborate pageant. With this in mind, I tried to avoid taking cute photos and instead presented the ritual as slightly unsettling. I photographed each animal as a spectacle in itself. Even a hedgehog was handled as if it were a precious object. King for a day.

During the ceremony of St Francis, the social constructs that go into dividing animals into subgroups – pets/livestock, cute/disgusting – partly dissolved. Such distinctions became irrelevant. Spectators were occasionally allowed to pet the animals assigned to various handlers. A teenage girl was cradling a brown rat in her arms. She gleefully observed the reaction of people as they first lightly stroked the friendly critter, then took a moment to realize that they had just petted a New York City rat. The subway kind. The look of kindness on faces was quickly replaced by horror, then confusion. In this context, a rat was just as cherished as the magnificent white horse, a crowd favourite. For one brief day, all animals present became equal.

The resulting photo feature ran in the *New York Times*. The reporter interviewed me for the piece and we discussed my existential musings about the ritual. In turn, he presented the question about animal souls to Bishop Daniel, who oversaw the ceremony. I was eager to get an answer from the legitimate authority. Unfortunately, the Bishop deflected, saying such questions were above his pay grade. In this theological and moral conundrum, I was once again left to my own devices. That night I had a cauliflower steak for dinner.

(above) Snowy owl, outside Cathedral of St John the Divine, New York, 2018
(overleaf) Tortoise and dromedary at the annual Blessing of the Animals, New York, 2018

w/ LARGE
ROW

INSIGHT

Tim Flach

The Animal Inside

TIM FLACH *is an animal photographer with an interest in the way humans shape animals and their meaning, while exploring the role of imagery in fostering an emotional connection. Flach is an Honorary Fellow of the Royal Photographic Society and was awarded an Honorary Doctorate from University of the Arts London.*

I AM INTERESTED IN how we depict animals to create pro-environmental outcomes. My commitment is to communicating stories. I create portraits to concentrate on character and personality. I have shot out in the wilds, but my expertise really is in describing the beauties of captive-bred animals, with images crafted in the studio. By taking the animal out of its natural context, I want to focus our attention on its innate beauty. Seeing the animal in this way takes us on a different kind of emotional journey. It's a reflection, a representation, a creative act that allows an inward exploration: what does that animal mean to you?

Take this Jacobin pigeon (opposite), for example. Domesticated birds. Did you know Darwin used to keep pigeons? He was obsessed with them and you can see a thread of pigeon thought running through his *Origin of Species*. So, I see pigeons and I think of Darwin. I appreciate that's not a connection many people make. The relationships between humans and animals are complex, we are all animals of course, but I mean there is complexity and conflict in interactions in the real world and the images of animals we create inside our heads. I think it's important to think more deeply about the thread of associations that encircles animals.

I also see my work as extending what many of the pioneering naturalists in the nineteenth century were setting out to achieve: to better understand and reveal nature through the best visual culture of the day. Take the artist John James Audubon for example. He saw, shot and stuffed specimens to make his magisterial natural history book – *Birds of America* is among the most highly-prized for collectors, sometimes selling for millions – but perhaps his marvels were not seen widely enough. Photography gives us the ability to reach audiences now that he could only have dreamed of. And keep the animals alive!

Images are naturally ambiguous. Even though most appear, at first sight, to be simply descriptive, images will always have more than one meaning. When you're dealing with people's experiences, it's really important to be interested in how they transform things into meaning and how that slips. How meanings ebb and flow between different groups of people. So when I'm working with a specimen or an animal, I'm considering what its symbolic meaning is. There are many animals that are famous as a character in fiction, or in a movie or an advert, but are little-known as a real species in the wild.

I play with iconography. Often referencing art or trying to untangle, to decode, some of the popular references to that animal within culture. All of these things, previously seen or recently made, act like a kind of cultural layer that wraps around the living animal. All of these things to some extent precondition the way each individual views an animal, whether in a zoo, a cartoon, or in a photo printed in a magazine, hung on a gallery wall or large format in a city

Red Splash, Jacobin pigeon, from the series *Birds*, 2018

Windows Chestnut, Arabian halter horse at the Ajman Stud, UAE, from the series *Equus*, 2008

‘Photography is powerful natural magic that can move people beyond the everyday ... It’s never been more important to connect people through nature; our future depends on it.’

centre. Some of my exhibitions have now been seen by millions of people. I get a lot of satisfaction from these big outdoor shows because the audience is so varied. Every person engages with a photograph in a different way, unique to them and the tapestry of their life experiences.

In animal portraits I look for character and personality. Anthropomorphism can act as an attention grabber, but my interests lie in what social scientists now call ‘critical anthropomorphism’. We can connect people to conservation stories and encourage pro-environmental outcomes. It’s a necessary ingredient for engagement. Research shows us that when people see romanticized images of a distant world, the kind you might enjoy in a traditional wildlife documentary or in the pages of a travel magazine, they are no more inclined to care about the animals. I have met many people on the edge of conservation who were surprised at the failure of romanticized wildlife images in communicating the causes they care about. I think the conservation movement, and wildlife film-making in particular, has often unwittingly created a world that separates humans and animals: over-dramatizing natural events and removing people from the frame. Mythologizing the wild.

I want my work to lean towards a different goal. The evidence shows overwhelmingly that when people connect to the character of an animal it evokes empathy. Some people have said my images are ‘uncannily human’. In other images we can see clearly the way humans have manipulated animals, whether in dog breeds or prize poultry, or other species now drastically endangered. I hope to encourage us to empathize with animals, to imagine their emotions and to reveal the animal in all of us.

In 2002 an Oxford study asked children to identify British wildlife images and then they asked them to do the same with Pokémon characters. Most of them could identify more Pokémon characters, so there is obviously a great need for us to encourage more understanding of the natural world. In my series *Equus*, horses were my focus; in *Dog Gods*, our canine companions. In *Endangered*, I traced the animal kingdom from the aquatic world through to the great apes. My gaze is always on the animal. My dog book went to fifteen editions in different languages, it was widely seen.

For another series, I looked at bats – a great symbolic subject. It’s always helpful to have something that’s got big guts to it. Cultural traction: you have Dracula, the bloodsucking vampire; you have Batman putting the fear of God into the criminal world. They’re laden with cultural superstition and they are great subjects to work with because I’m fascinated by that junction between culture and the natural world.

Aware of cultural associations, you can play with memories of a famous picture, ambiguities and changing attitudes. I photograph a horse and think of a Stubbs oil painting, but also lean in to aspects of modern horse culture, training or animal husbandry that only experts might appreciate. When we make images I think we are constantly moving between multiple interpretations. There is slippage and change. We must be interested in other people’s perspectives. The same animal can mean dramatically different things in different countries.

Here’s a behind-the-scenes image of a chicken (p. 171). I didn’t pluck it for the shot,

Axolotl, photographed in Flach's London studio, from the series *More than Human*, 2012

Featherless chicken, photographed at the Hebrew University
of Jerusalem, from the series *More than Human*, 2012

that's important to say, it's a featherless chicken. We were at an agricultural college in Israel where they're breeding them for the developing world so that they don't have to keep them so cool. What's interesting for me about this picture, and I know it's a bit disturbing, is that, as we look, we imagine it presenting for our gaze, as if it's a kind of plump ballerina moving across a stage. It's looking back at us and if you think about it, most of us – certainly me anyway – don't keep chickens any more. Our only experience of them is dismembered in cellophane in the supermarket. And that's how millions of us meet animals today. Something has gone wrong, right?

* * *

I often rely on cuteness as it's a powerful tool. When it comes to communicating, we must understand that the symbolism of animals is as important as an understanding of the biology or zoology of the species. Whether it's a red cardinal that looks like Angry Birds, or a pied tamarin from the Amazon that could be seen as a Yoda, these are not facile cultural connections. They're meaningful. It's what people bring to the animal, to the photo, when they meet it for the first time.

When I photograph an axolotl, I want to give it a character that its predicament demands. It's a type of salamander, in case you don't know, and it's critically endangered. I've photographed a species in Mexico that is now almost extinct in the wild. There are layers of Mayan indigenous folklore surrounding it, as well as storylines of modern science. Those aren't feathers on its head, but unique gills that help it breathe underwater. It has amazing healing powers: if it loses a leg, it can grow a new one. Another aquatic salamander, the olm, can be found in caves in Croatia. It's almost totally blind and lives in near darkness. It survived the impact of the asteroid that brought the end to the dinosaurs. Tell me these creatures are not incredible!

As photographers, we can give all kinds of animals our time, in the hope others will do the same. I want people to look for more than ten seconds – that's how long researchers have found people spend, on average, with artworks in a museum or gallery. Animals need our awareness of the realities of their lives. From beloved pandas to charismatic tigers, to the little-known saiga antelope, tree frogs or rare salamanders. It's not anthropomorphism in the crude, old-fashioned sense. It's a desire to align sameness with otherness. Photography is the method I can use best to inspire a connection between human and non-human animals.

I believe very strongly that we need to use the arts to take science and shift it into popular culture. That's how people's views can change. With the right stories, empathy can move people to action. Like this, I do believe – and you might think me naive and too optimistic – that photography can bring animals back from the edge of extinction. It's easy to be negative. Far harder in today's world to search for the positive. Denial and doom are prohibitive realities. We must use the aesthetics of wonder.

Yellow-eyed tree frog eggs, photographed at Manchester Museum's Vivarium, from the series *Endangered*, 2017

Dalmatian puppies, from the series *Dog Gods*, 2010

Stefan Christmann

STEFAN CHRISTMANN *is an award-winning nature photographer and film-maker from Germany. In Antarctica he documented the lives of emperor penguins from their birth until adulthood as part of a three-person film team for the 2018 BBC programme* Dynasties. *His photos from that assignment won the Portfolio Award at Wildlife Photographer of the Year, among many other honours.*

Is there a particular photo, or group of images, that inspired you into action? A Canadian goose protecting two of her chicks on a snowy day on the cover of the magazine *Nature's Best Photography*. It was an image that touched me on an emotional level and it showed that even though an image was not technically perfect (which is something I thought was always needed in winning images) it could still convey emotion and a sense of wonder. Up to this day, it is one of my favourite photographs.

Is there a particular photo, or body of work, that raised your curiosity like no other? Really tough. I loved Hans Strand's body of work *Iceland from Above* and Art Wolfe's book *Edge of the Earth, Corner of the Sky*. I remember that when I had Art sign my very loved and slightly battered version of the book, he chuckled and told me that he really liked the fact that this book was being used and not just sitting on a shelf.

What is the most influential animal photo ever taken? I would say a lot of the Wildlife Photographer of the Year winners each year tick that box. But for me personally, it would be John McColgan's shot of two mountain elk taking refuge in the Bitterroot River as a fire rages around them and Justin Hofman's shot of a seahorse swimming with a discarded cotton swab. Such impressive, thought-provoking images and taken in a moment.

So, why do you photograph animals? They are out of my control, and I enjoy their presence and company. I also do some product photography where I can control virtually anything from light to subject – also fun, but not as calming. Interacting or being allowed into an animal's vicinity gives me an incredible amount of satisfaction. To me all animals are somewhat innocent. They don't have any hidden agendas and their motives are pure. That reflects back on my own life and it just calms me down immensely.

Are animal photos important? I think there are basically two kinds of images: photos that show the beauty and photos that show the bad things. Both are equally important. The 'bad' photos show us how we humans do not take enough care of the planet and how we do not treat the beings amongst us with enough respect and dignity. Those images make us question how we act as humans. Sometimes these images can convey a feeling of 'all is lost' and that is exactly why we also need beautiful images. We need to show people that there is still beauty in this world and that there are still places that need protection. We need hope.

But, you know, even a cat meme or a pet video can have a profound impact on people. I follow an Instagram channel created for a pet cockatoo named Fefe, who grew up in captivity and started plucking her feathers. She was adopted from an animal shelter by a couple who post videos of her life. We humans can be caring and it proves that living in harmony with animals is fulfilling. It also shows that even though an animal was in distress, life can always change for the better. Their cockatoo photos will never win WPOTY, but they still inspire people for the better. To me this is just as important.

Emperor penguin chick hatching, Atka Bay, Antarctica, 2017

‘I’ve only been photographing for twenty years but I’ve already collected a lifetime of wonderful memories. Photography helps me evolve as a human being.’

Stefan Christmann

(above) Emperor penguins huddle against the cold, Atka Bay, Antarctica, 2017
(opposite) Emperor penguin and chick, Atka Bay, Antarctica, 2017

Leila Jeffreys

LEILA JEFFREYS is an Australian photographic and video artist who began documenting birds by way of photographic portraiture in 2008, working alongside conservationists, ornithologists and bird sanctuaries. She has published three photographic art books. Her work has been exhibited in Sydney, New York, Hong Kong, Stockholm, Brussels and Istanbul, and in 2023 was included in the Civilization *exhibition at London's Saatchi Gallery.*

What are your most-viewed animal photos? I find that people have a very deep connection to cockatoos. On the east coast of Australia we have wonderful yellow-tailed black cockatoos. They are huge, with a beautifully slow, lolloping flight and a prehistoric call. My portrait of a yellow-tailed (but really any cockatoo) seems to hit people in the heart. They are the birds that people easily fall in love with. My portrait of Tani, a masked owl, would be another, but then also my portrait of a rose-crowned fruit dove. I think people couldn't believe that a dove (aka a pigeon) could be so beautiful.

Is there a particular photo, or body of work, that raised your curiosity like no other? I love the work of Finnish photographer Pentti Sammallahti and Nick Brandt's *Inherit the Dust*. Nick shot some haunting images of calcified birds in Tanzania's Lake Natron. The waters in the lake are highly alkaline, killing unlucky animals and seemingly turning them into stone. Those photos had me looking up and reading everything I could about it.

What do you hope to achieve? When I started taking pictures, my work was about expressing all the ways in which birds are extraordinary. I wanted to direct the viewers' attention to the jaunty shape of a cockatoo's plumage or the curious expression that flits across an owl's face. But the more I've grown as a photographer, the deeper my desire to awaken the viewer to the powerful web of connections we share with the flying creatures we take for granted. I want to expose the kinship that exists between the human and non-human.

Is there an animal photo that you most frequently think about? Paul Nicklen's photo *Suspended Grace*, which is of sleeping sperm whales. I love the sea and that photograph encapsulates the sense of vastness and beauty of the oceans, as well as an incredible sense of stillness and peacefulness when we sleep. Sleep is a shared experience with all animals on this planet.

Can you describe the positive impact that one of your photos has had? I receive messages of support from many people who have said they never noticed birds before and do now. That they went from being blind to them to really 'seeing them', and that then led to joy in discovering them and learning more and more about the birds in their area. From there they have thought about how their actions can have a direct impact on birds' lives. It might take a while, but yes, I am hopeful that photographs do have the power to shape more positive behaviours.

Are animal photos important? Photography is a chance to show respect for the things someone loves. People photograph animals for all kinds of reasons, but for me it's about celebrating the wonder of the natural world and trying to rekindle inside each person a love for the animals we share our world with. I hope that my art practice further develops that connection in me, and then I hope that 'energy' sparks an inner change in others. It is sometimes said that 'thousands of candles can be lit from a single candle and the life of the candle will not be shortened'. Happiness never decreases by being shared, it only grows.

The Tweets, a trio of budgerigars, Canberra, Australia, 2018

‘Photography is about reconnection. We have become so disconnected from the natural world. This reconnection is something that I’m searching for, for myself and for others, because I feel like it is one way to solve a lot of the suffering on the planet.’

Leila Jeffreys

(above, left) Rose-crowned fruit dove, Taronga Zoo, Sydney, 2016
(above, right) Orange-headed Gouldian finch, Finch Society of New South Wales, Sydney, 2013
(opposite) Melba, a yellow-tailed black cockatoo, Kaarakin Black Cockatoo Conservation Centre, Perth, Australia, 2011

Sirocco, kakapo, Wellington, New Zealand, 2015

Tani, Australian masked owl, Queensland, 2013

Anuar Patjane

Anuar Patjane is a Mexican social anthropologist, photographer and scuba diver. He is a regular contributor to National Geographic *media and many environmental and photography magazines. He is a World Press Photo winner in the 2016 nature category, a* National Geographic Traveller *photo contest winner and a winner at the Cannes Lions creative awards.*

Who are the wildlife photographers that you most respect? A master photographer, Sebastião Salgado, who has turned to nature photography in recent decades. These are shots of raw power and pure beauty. In the more conventional sense of wildlife photography, Paul Nicklen and Tim Laman are the best in the business. They go the extra mile to secure unforgettable shots.

What is the most influential animal photo ever taken? There is footage of an orangutan fighting away a bulldozer. His jungle home has been destroyed and the future for his species is bleak. That sequence is iconic and sadly so representative of our times.

So, why do you photograph animals? I do not see myself as a wildlife photographer or a photographer of animals. I photograph humans *interacting* with animals. To be brutally honest, I find most conventional wildlife photography pretty dull. There needs to be more poetry, more curiosity.

What do you hope to achieve? To create emphatic feelings and connection with nature so we, the dumb monkeys, can connect with other animals and stop abusing them. We don't need to fear other animals, they just need our respect. In a phrase: I want my photographs to transmit coexistence.

Is there an animal photo that you most frequently think about? Kevin Carter's shot of a child and vulture in Sudan is unbelievably sad and has been much written about since, but you can't deny its power. Totally different in intention is the work by Philippe Halsman, *Dalí Atomicus*, which shows the artist leaping in mid-air, with an arrangement of paintings, water and three cats flying in all directions. I've read it took them twenty-eight goes to achieve it. It's a kind of poetry, I think, and very playful. I do prefer photographs in which humans interact with animals, or at least images that are more than just a straight shot.

Can you describe the positive impact that one of your photos has had? Once in a while I get an email from people telling me that, thanks to a photograph that they saw somewhere, they began scuba diving. Some mention that the photographs helped them lose their fear of underwater animals or the fear of not knowing what is beneath them in the water. When I get these kinds of emails and comments, for me that's it! If you begin to dive, you will love the ocean and that's the very first step towards sea protection and sea regulation.

Some of my happiest and most memorable moments come from diving and, for example, unexpectedly finding a giant humpback whale helping her calf with the breathing technique, while dolphins play around them. That is a kind of heaven for me.

It is often said that 'photography can change the world'. Do you agree? Yes it can, but it is difficult to measure. All art forms have this power – music, cinema, photography – and the capacity to change the course of action by forcing us to think differently. How many wars have ended thanks to the change in perspective of public opinion. And what makes people change their minds? Photography and, now, video too.

(above) School of bigeye trevally, Cabo Pulmo, Mexico, 2017
(overleaf) A humpback whale and her calf (and photographers) near Roca Partida, off the Pacific coast of Mexico, 2015

Park ranger and a school of bigeye trevally, Cabo Pulmo, Mexico, 2015

PROFILE

John Bozinov

JOHN BOZINOV is a New Zealand-born photographer and educator. He first voyaged to Antarctica as an Enderby Trust scholar and has since worked there regularly on expedition ships. He is on a mission to help protect our wilderness areas, one iPhone shot at a time.

Can you remember the first photo of an animal that you took? I've always been interested in taking photos of animals, but wildlife isn't an easy subject when you're starting out. I used to take hundreds of photos of my cat, following him around the garden or sitting on the couch. He was my muse as I learned to use my camera. Practise on your pets – and if you don't have a pet, well, make friends with someone who does!

What is the most influential animal photo ever taken? There was a photograph a few years back of an emaciated-looking polar bear walking on ice, seemingly on the verge of death. It was photographed and filmed by Paul Nicklen and Cristina Mittermeier. Despite its controversies, I think this photo has stuck with people as a symbol of climate change and the effects of our warming planet. It's a challenging photo.

So, why do you photograph animals? I enjoy the limitations and the challenge of photographing a wild animal in its natural habitat. I can't direct it, change its posture or adjust the light: it's just me and my camera making the most of what's out there. Limitation is and always will be the soul of photography.

Has an animal photo forced you to examine your life? I recently saw a photo of Mayombe, a western lowland gorilla. She's the first captive-born gorilla to be released and then successfully give birth to a baby in the wild. The look on her face as she holds her newborn is so human-like that it completely changed the way I thought about non-human primates. A photograph can tell this story and elevate the issues. It's a huge step for global conservation.

Is there an animal photo that you see as your stand-out shot so far? For polar photography my favourite camera to use is the iPhone, though it's limited in that it doesn't have a long telephoto lens. The shots that stand out to me are those that are particularly difficult to capture on the iPhone's standard wide-angle lens. I recently made some shots of newborn penguin chicks with their mother, fairly challenging to get without disturbing them, so I'm super stoked with those.

Are animal photos important? Somebody once told me they think zoos are important because they give people an opportunity to see animals that they'd never have the chance to see in the wild. I don't think anybody has some essential right to see a particular animal. For the most part we should probably limit animals' exposure to humans and the impact we have on the planet. Conversely, people are visual animals, we usually need to see to understand and appreciate things. Photos of animals, especially those we don't often see or hear much about, can help us connect with these species and value their preservation and protection.

Fur seal, Deception Island, Antarctica, 2018

Kea, Arthur's Pass National Park, New Zealand, 2014

Gentoo penguin, Deception Island, Antarctica, 2016

Hengifoss sheep, Iceland, 2017

Atlantic puffin, Saltee Islands, Ireland, 2019

Claire Rosen

Fantastical Feasts

CLAIRE ROSEN *is an artist whose elaborate constructions often feature whimsical anthropomorphic animals or symbolic still lifes, evoking the aesthetics of classical painting. Rosen has been named twice to the* Forbes *'30 under 30' list for Art and Design and recognized by Communication Arts, IPA, Photolucida and Prix de la Photographie.*

THROWING ELABORATE DINNER parties for all manner of animals, from honeybees to elephants, is one of the things I love most about being an artist. My series *Fantastical Feasts* uses the icons of the banquet and Last Supper to subtly encourage viewers to consider those in the animal kingdom from a different perspective, reflecting on their social lives and what rights they might have. To see animals humanely sometimes requires us to see them as human-like. By placing animals in a setting typically reserved for humans, it raises the question of whether we may have more in common than we admit. The feasts are a whimsical invitation to the viewer to reflect on the nature of society and our relationship to and responsibility towards the creatures we share the planet with.

From an early age, I was captivated by creatures big and small, real and imagined. From the charming adventures of small woodland creatures in their best outfits illustrated by Beatrix Potter, to thrilling encounters with animal ambassadors at local zoos and farms, to hours spent entranced by the taxidermy animal vignettes at the Museum of Natural History in New York. Though I didn't know it at the time, this parade of childhood animals was building a vocabulary of surprise and delight that would later become the language of my artwork.

An artist residency rekindled my childhood fascination with natural history and I painstakingly photographed hundreds of vintage specimens. I was inspired by historic illustrators of flora and fauna like John James Audubon and Maria Sibylla Merian, assemblage artist Joseph Cornell and contemporary installation artist Mark Dion. This project led to a commercial collaboration with a chandelier designer who incorporated ethical taxidermy into some of her designs, which we photographed in scenes with live animals. Included was a scene of pigs enjoying a table of doughnuts under large, gold chandeliers that would become the inspiration for my *Fantastical Feasts* series several years later. I came to appreciate just how much I enjoyed working with live animals. It was thrilling. I was on a commercial fashion and advertising path at the time, but any chance that I had I would incorporate an animal into the concept.

When I finally focused my attention fully on animals in my series *Birds of a Feather*, it was a turning point for me. I photographed live parrots against historically inspired wallpaper selected for beauty and visual blending. I was captivated by the unique expressions of the birds, their plumage and beaks, and the unpredictable nature of the shoot. To my surprise, the series saw some success. This gave me the confidence to abandon the fashion and advertising imagery I had believed I needed to be commercially successful to fully pursue fine-art photography. In addition to attracting positive attention, some press coverage of the series also attracted

The Cobra Feast, Jaipur, India, 2016

‘Photography is a magic art that can awaken, compel, inspire and confuse. It can astonish and infuriate. And, often, many of these things at the same time. Photography encourages us to really feel.’

negative feedback from animal rights activists who were critical of the pet industry.

This was an eye-opening experience for me, and the commentary led me to investigate the issues facing these feathered creatures. Parrots are intelligent animals that need proper care and companionship or their mental state deteriorates. Far too many birds are surrendered to shelters because their owners are not able to care for them. This awareness shifted the meaning of this project for me. I began to see the wallpaper as a symbol for the man-made interiors these birds inhabit when we acquire them as pets. The birds are beautiful and appear to belong, when in reality they are a far cry from their natural environment. I began donating a small portion of my print sales to rescue organizations and realized that my work could be more than just decorative, that it could raise funds and create conversations. Though the first conversation I had started with that series had not been kind, it forced me to re-examine how I engage with animals in my life and through my work and the dialogue I wanted that work to foster. I went on to create several offshoots of this concept with owls and raptors, ducks and chickens, snakes and insects.

* * *

The *Fantastical Feasts* project is a series of constructed images featuring live animals eating around elaborate banquet tables. I set tables all over the world and invite animals of many stripes, spots and scales to come and eat. It is both exhilarating and nerve-racking to work on these shoots. It starts with imagining what the centrepieces at a cobra party would look like, or what wolves might enjoy eating – then the scavenger hunt at local markets begins. I arrange everything as meticulously as I can and then completely surrender control to the subjects, waiting with bated breath to see what they do when they get to the table.

The magic of the shoots is in their unpredictability. And at times they fail completely! After setting a table for Andean condors in Peru, I waited an entire day as they sat staring at me from their perches, never touching the banquet I had prepared for them. I am also cautious in my work. While shooting the hyena feast, I was tucked safely away inside a cage, with caretakers on either side of me, while the food and some of the table settings were enthusiastically devoured. The comfort and safety of the animals has the highest priority, and caretakers are always close at hand. All the food and arrangements are approved in advance. For the most part, the animals are far more interested with what’s on the table and they don’t seem to care about me or the camera at all.

Of the *Fantastical Feasts* series, *The Sloth Bear Feast* is perhaps one of my favourites. The bears had made their home at Wildlife SOS, a rescue facility in Agra, India. The organization had done incredible work to educate and transform those who used to exploit the bears for income in the tourism industry into guardians of those same creatures. Prior to the shoot, I had been advised by their caretakers of their favourite treats and the bears were delighted by the honey and watermelon that they discovered once they were invited to roam the banquet table.

The Cobra Feast is another favourite for a different reason, inspired by the impulse to encourage people to see the beauty in a species

Spotted Eagle Owl No. 7261, from the series *Birds of Prey*, Stellenbosch, South Africa, 2016

The Sloth Bear Feast, Wildlife SOS Bear Rescue, Agra, India, 2018

they might otherwise fear. Snakes are persecuted the world over. While some species are very dangerous when threatened, these king cobras particularly so, they are also incredibly beautiful and play a critical role in the ecosystem. One of the joys I get from this series is seeing how viewers react to the images, and the cobras never fail to provoke a visceral reaction, either a drawing into or a physical recoiling from the picture. I'm both frightened *and* fascinated by these creatures, and they inspire me with the ancient folklore and mysticism that swirls around them.

While my work differs from that of many contemporary wildlife documentary photographers, I hope to contribute to a mindset shift and a deeper conversation around some of the nuanced issues facing animals in the world. Through my work I have experienced and seen the deep care and connection that we can have for non-human creatures of all types, and how our empathy can extend to the most unusual critters. Modern society has made it incredibly difficult for us to practise such care in our daily lives and almost impossible to avoid doing harm to creatures that we would never wish to hurt. The chain of events that is connected to the purchases we make and the way we live are often invisible to us, their impact hidden from our view.

I hope for us all to be more mindful of our actions and aware of the consequences of our choices, always thinking of how we can collectively lessen our negative effects and be more creative and resourceful as we seek solutions. I believe that only the power of our collective longing for a different approach to animal interactions, testing, food systems and land-use can usher in the transformation we need.

Sometimes photography that raises awareness can be hard to look at and you don't want to hold on to all that anguish in your mind. I feel that one of the opportunities offered by the *Fantastical Feasts* project, in all its whimsy and colour, is that the images stay with you for a while and engage a different part of your brain. The message tries not to accuse, but rather to engage and spark the imagination and action of the viewer.

How can we arrive without blame at a solution to the problems we all now face? How can we change the exploitative relationship we've had with animals in the past and learn to protect our coexistence with them? I hope my photographs can be part of conversations about how we can do things better. And I do believe that we can always do things better, from our smallest daily decisions to our greatest aspirations as to how to live mindfully in this entangled world.

Flamingos No. 0365, from the series *Birds of a Feather*, Miami, Florida, 2017

Fragments

Evgenia Arbugaeva, walrus outside the haul-out hut, Chukotka, Russia, from the series *Hyperborea*, 2018

Morgan Heim, yearling black-tailed deer, Oregon, from the series *A Last Leap Towards Flowers*, 2017

Victor Prout, 'Native tiger of Tasmania shot by Weaver', Hobart, New Zealand, 1869

Maroesjka Lavigne, *White Rhino*, Namibia, 2015

Men posing with a mountain of bison skulls at a glue works, Rougeville, Michigan, 1892

Arthur Rothstein, sheep awaiting shipment, Belfield, North Dakota, 1936

Mitsuaki Iwago, blue wildebeest and Burchell's zebra migrating, Serengeti, 1987

Andreas Gursky, *Greeley*, a cattle feedlot in north-eastern Colorado, 2002

Keystone View Company, '3,000 sheep astray on a mountain range', Montana, 1900

George Shiras, white-tailed deer caught in a 'flash-trap', Michigan, 1893

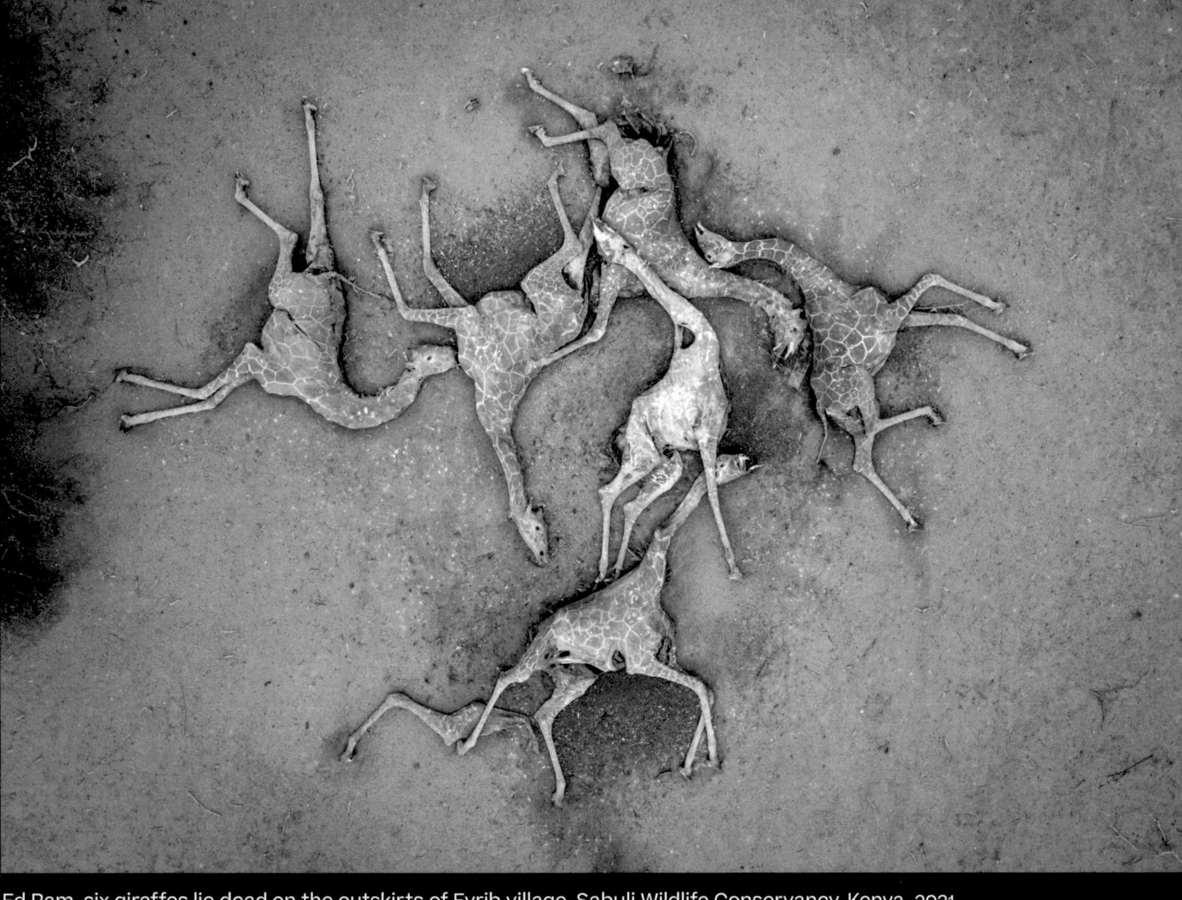

Ed Ram, six giraffes lie dead on the outskirts of Eyrib village, Sabuli Wildlife Conservancy, Kenya, 2021

Frederick William Bond, contents of an ostrich's stomach, London Zoo, 1927

Chris Jordan, fledgling Laysan albatross, dead after ingesting plastic, Midway Atoll, 2009

Daniel Beltrá, brown pelicans covered in crude oil, 2010

Brian Skerry, bigeye thresher shark trapped in a gillnet off the coast of Baja California, 2010

Marcin Ryczek

MARCIN RYCZEK *is a Polish-born fine artist whose photography often refers to symbols and geometry. Its characteristic features are simplicity, minimalism and reference to graphic arts.*

Do you have a favourite animal photo that you have taken? All photos that I create of animals are important to me, because each of them has a story that is intimate to its creation. Certainly, such a photograph is *A man feeding swans in the snow*, which is now one of my most recognizable images. This is a very personal and real photo. I made it at a time in my life when, after some difficulties, I felt calm and balanced. I had many thoughts, that sometimes seemingly negative experiences can lead us to something good. I wanted to picture it and express what I feel with photography.

So, why do you photograph animals? Harmony. In my work I often refer to nature because I believe that we are a part of it and being close to it, caring for it, is an indispensable need for most of us to feel balance in our lives....Photography and observation of the animal world is a kind of meditation.

Can you describe the positive impact that one of your photos has had? It's incredible to me how many beautiful messages I've had from people all over the world inspired by that shot of swans in the snow. Each person interprets it in their own way. I was moved by the letter of a lady who thanked me for this photo, saying that her husband, who has schizophrenia, brings up my photo on his computer in difficult moments and it relaxes him and puts him in a better state. I have other messages from people who found that the photo helped them overcome some weakness, solve their problems, or gave them hope that beyond the dark side there is also the light side. There was another message from a girl from Spain, who wrote that she had been functioning with a sore tooth for a long time and, despite the pain, was afraid to visit the dentist. But on the day she saw this photo, it influenced her and encouraged her to get an appointment, and she finally rid herself of the pain in her teeth! As I said, surprising stories are born from a single image.

It is often said that 'photography can change the world'. Do you agree? Yes, I believe that photographs can have a significant impact on our lives and the world. It's hard for me not to believe it when people take the time to write to tell me how an image I've created might have changed their own life in some way. It's very humbling. But, yes, more famous photos can certainly stimulate thinking, changes in lifestyle, reflection on the injustice of our world, or greater care for our planet.

Photos are also used to manipulate, to bend the truth. Sometimes, even when serving a good cause, people show photos that were not taken in a given place or time. It is not verified on the Internet and people believe and share unverified content on Facebook or Instagram on a large scale. I think that in photography, especially journalistic photography, it is worth reaching for the truth, and this truth about the world in photography may provoke change and improvement in the world.

Do you have any other thoughts to share?
I remember a scene in a Polish film, where a soloist who sang in a choir was asked by a journalist after the concert, 'why sing?' He replied: 'I sing because then I feel like a better person.' I think that can also apply to photography. Photography sensitizes. When I go out with my camera, I'm more open to the beauty that surrounds me. Communing with nature, animals and people makes me a better person for that moment.

(above) A man feeding swans in the snow, 2013

(overleaf) Liberation, 2014

PROFILE

Nichole Sobecki

NICHOLE SOBECKI *is a photographer and film-maker based in Nairobi, Kenya. Her work has been recognized by the Leica Oskar Barnack Award, the ASME Award, the Robert F. Kennedy Human Rights Award and Pictures of the Year, and has also been exhibited internationally.*

Can you remember the first photo of an animal that you took? I rode and trained horses and worked at a barn near my house, so my first animal images were all equestrian.

Who are the wildlife photographers that you most respect? I'm most moved by photographs that help me to see the natural world in new and unexpected ways: Xavi Bou's bird flight composites, the dreamy quality of Evgenia Arbugaeva's work, which is full of animals if not typically focused on them, Maggie Steber's *Dead Lizards Army*, Claire Rosen's imaginative *Fantastical Feasts* series or Anand Varma's work on hummingbirds.

Do you have a favourite animal photo that you have taken? A hawk taking flight in the ancient port city of Maydh, Somaliland, which was quiet and nearly abandoned when I visited, because for me this image has come to represent the future awaiting this region if the worst effects of the climate crisis are not mitigated. And also the image of a seven-month-old cheetah, later named Astur, in the back of a Land Cruiser hissing at a rescuer's outstretched hand.

Is there one historic shot that you really admire? Eadweard Muybridge's galloping horse with rider. In 1872 a racehorse owner hired Muybridge to settle a bet about whether galloping horses' hooves were all fully off the ground at the same time – Muybridge's images would go on to prove that they were. It's a reminder of how the camera can remake our understanding of the world and help us to see beyond our own limitations (physical in this case, but that idea can be extended to cultural or ideological barriers too).

Can you describe the positive impact that one of your photos has had? In 2019 writer Rachel Fobar and I visited Pienika Farm in Lichtenburg, South Africa, where captive lions are bred and which also offers sports hunts. Our story followed a visit by the NSPCA [National Council of Societies for Prevention of Cruelty to Animals] that their senior inspector described as 'soul destroying'. After a long, committed campaign, in May 2021 South Africa finally announced plans to clamp down on breeding lions for hunting or so tourists can pet cubs. This was a collective effort by many individuals and groups, but I felt really humbled to be a part of helping to raise awareness about this harmful practice. There's no formula for measuring a picture's impact, but this was a case where I felt a discernible shift.

Are animal photos important? Animal photographs can reinforce our connection to the natural world and break down the false pride in our separateness that I believe enables the horrors we perpetrate against nature, animals and one another.

Do you have any other thoughts to share? I think that one hopeful thing I see in art and culture these days is the growing number of people trying to challenge this idea of human dominance over the natural world. We need new forms of knowledge and also the return to old, indigenous understandings. We're relearning how to see the non-human world as more than simply a resource for the taking, but as deeply alive, sentient, even sacred. I think arts are helping to carve a path towards a societal transformation where this realm of thinking becomes more and more commonly understood.

A hawk takes flight in the ancient port town of Maydh, Somaliland, 2016

A seven-month-old cheetah in the back of an SUV hisses at a rescuer's outstretched hand, western Somaliland, 2020

Five rescued cheetah cubs keep warm in a tent,
Hargeysa, Somaliland, 2020

Lobikito Leparselu cares for his family's goats, northern Kenya, 2020

Marsel van Oosten

MARSEL VAN OOSTEN *is a Dutch photographer specializing in nature and wildlife photography. He has been overall winner of the Wildlife Photographer of the Year and Travel Photographer of the Year competitions.*

Do you have a favourite animal photo that you have taken? It's an image of an elephant standing at the edge of Victoria Falls in Zambia. It's the only photograph in the world with an elephant at the Falls, an extremely rare sighting. The image also contains all the elements that I'm after in my photography style.

Is there a particular photo, or body of work, that raised your curiosity like no other? The first time I saw images of Japanese macaques bathing in natural hot springs in Japan I was absolutely fascinated. Their behaviour was so human-like.

What is the most influential animal photo ever taken? Instead of singling out just one shot, I would like to say all images that show the brutality of poaching and the illegal wildlife trade. One of the many great photos by Brent Stirton comes to mind – a dead rhino, killed by poachers, horn removed. It's a graphic shot, very in your face, but the way it was photographed with flash made it look a little beautiful too. A bit surreal and definitely uncomfortable. Although I don't enjoy looking at images like that, they are very important for addressing the threats to wildlife around the world faces and for creating a sense of urgency.

So, why do you photograph animals? Because I love animals. I love nature. And I love being outdoors. Animals are fascinating, beautiful and there is such a huge variety. The most frustrating part about wildlife photography – the fact that you have virtually no control over your subjects – is also what makes it so interesting. Wildlife is inherently unpredictable. You never know if and where animals will show up and, if they do, you never know what will happen. This makes it very exciting. No single day in the bush is the same. In wild nature most things are the way they are and have been for aeons. It's life in its purest form. The human world is the opposite – a world of hidden agendas and make-believe. People always seem to focus on trivial matters: the latest fashion, the latest outrage. In nature, it's all about the essence. I'm a bit of a misanthrope.

Has an animal photo forced you to examine your life? There is a photo that shows me running for my life while being chased by a tiger. I've always been convinced that we only live once, but that picture is the living proof that you should not waste any time on trivial stuff that does not make you happy – life can be over in the blink of an eye, or the reach of a claw.

It is often said that 'photography can change the world'. Do you agree? Photos can inform, entertain, influence, scare, amaze, disgust and trick people. It's an extremely powerful medium. When I worked in advertising, I used imagery to attract people and seduce them. It can be used for good and for bad. As a wildlife photographer, I use my images to inspire and awe people. I hope that by showing the incredible biodiversity our planet has to offer and the beauty and importance of precious habitats, I can reconnect people with nature and make them realize that we have a lot to lose if we don't look after it. I know many other photographers say this too. The more the better. We have a lot of planet to protect.

Are animal photos important? It all depends on the objectives. I'm an artist first and a conservationist second. My primary objective is to create a thing of beauty – my personal aesthetic interpretation of our natural world. I then use the result to inspire and educate people. I would like to think that that is important.

African elephant at the edge of Victoria Falls, Zambia, 2007

(above) Golden snub-nosed monkeys, Qinling Mountains, China, 2016

(overleaf) Oryx running in the dunes, Namib-Naukluft National Park, Namibia, 2009

PROFILE

Staffan Widstrand

STAFFAN WIDSTRAND *is a multiple award-winning nature photographer based in Sweden. The author of eighteen books across nine languages, he works to highlight the attractiveness of natural heritage and inspire wide audiences to better protect and take care of it.* Outdoor Photography *named him one of the most influential nature photographers in the world.*

Can you remember the first photo of an animal that you took? Weaver birds nesting, when I was seven, in Uganda with my dad's camera. The second shot I vividly remember with my own camera, a Kodak Instamatic, when I was twelve. This time on a family visit to Kenya. Elephants! I also managed to get a shot of a leopard in a candelabra euphorbia tree. Looking back now, these early experiences were really important for me.

Describe an animal photo you've taken that was memorable to you. Pallas's cat and brown bears, both because of the close distance and the eye contact. Had the cat been, let's say, a tiger or a lion, that would have been a very scary encounter indeed, looking at me in the way the predator looks at the prey. The bear is the opposite. It also looks at me, but with kind eyes. Rather, the bear seems to be saying: 'please don't harm me and I promise that I won't harm you either.' It was less than two metres away, with me in a canvas hide. It could have walked straight through the tent. But it didn't. We had a quite relaxed mutual win-win understanding somehow.

Is there one historic shot that you really admire? An orangutan swinging on a vine by Frans Lanting. It was an innovative photojournalistic shot at the time, with motion blur, sharpness, fill-in flash. The man with a moa egg in his hands by Frans Lanting. The puma in front of the Hollywood sign by Steve Winter. Cassowary images by Christian Ziegler.

What is the most influential animal photo ever taken? Maybe Jim Brandenburg's white wolf jumping between ice floes on Ellesmere Island. Petrels in front of a blue iceberg by Cherry Alexander. Several of Britta Jaschinski's images from the wildlife trade. And Brent Stirton's mountain gorilla, killed by poachers, being carried down from the jungle. It looks as if the gorilla has been crucified. It's a very moving and iconic image.

What do you hope to achieve? To inspire people to fall in love with the natural world. Any and all parts of it. What you love, you will protect, so we need to make more people fall in love with our natural heritage.

Can you describe the positive impact that one of your photos has had? I have been part of major, multi-year projects: *The Big Five*, *Wild Wonders of Europe*, *Wild Wonders of China*, *Rewilding Europe*. These were long-term, mass-communication projects with reportage, books, magazine stories, kids' books, social media, huge touring outdoor and indoor exhibitions and TV, reaching several dozens of millions of people. *Wild Wonders of Europe* was estimated to have reached at least 800 million people.

It is often said that 'photography can change the world'. Do you agree? Yes, definitely. Photography can inspire and make minds open up for additional communication. Same with film-making. Images help change perceptions. Images stimulate empathy. Yes, images can help save a species.

Are animal photos important? They help a wide audience see the beauty, the aesthetics, the cuteness, the freedom, the soul of the natural world and the incredible wild beings that populate it.

Pallas's cat, Qinghai, China, 2016

Gaoligongshan tree frog, Yunnan, China, 2012

Bearded seal, Spitsbergen, Svalbard, 2014

Red panda, Laba He nature reserve, Sichuan, China, 2016

Alicia Rius

ALICIA RIUS *is an animal portrait and lifestyle photographer. Whether on location or in-studio, Alicia creates artistic portraits with well-crafted light and composition that draw the viewer in to reveal her subject's spirit and personality. Her animal work has received an IPA Award.*

Can you remember the first photo of an animal that you took? It was my Persian cat Justina. The very first photos I took with an analog camera. She had just been groomed and, for some reason, the groomer thought it would be fun to make her look like a lion. My mom was super upset and Justina probably a little traumatized. I thought it was pretty funny and she inspired me to create a little photo story about a lion that attacked me. I took the photos, sent them for developing and then stuck them in an album when they came back. I took the album into school for kids to see and people enjoyed it.

Is there one historic shot that you really admire? I adore the work of John Drysdale. His collection *Our Peaceable Kingdom* is filled with marvellous shots of exotic pets, my favourite being *Descending a Staircase*. I love this photo because it makes me reflect. Technically, it's perfect – the composition, the moment he caught, the expression on the girl's face, the alligator's look in its eye – and yet, ethically, it's just wrong on so many levels. An alligator should never be a pet and it can never be tamed as a pet, let alone letting a toddler play with it. But that was 1976!

Is there a particular photo, or group of images, that inspired you into action? Yes. A photo of a cat with a birthday hat and sunglasses. In fact, any shot where a pet is made to look ridiculous with human accessories. When I moved to San Francisco in 2013 and decided to become a pet photographer, I did my research. I remember seeing all these cheesy photos for postcards and I knew I could do something different. And had to! That's how it all started. With a simple mission to elevate an animal's natural beauty, with no props.

So, why do you photograph animals? The question should really be 'why not?' If we spend our lives documenting our kids, our weddings, our trips and parties, why wouldn't we want to capture the lives of our companion animals? The most loyal and forgiving creatures that exist. Loved like our own children, they make our days better and bring out the best in us. Why wouldn't you want to capture that? Why wouldn't you want to immortalize that bond you have with your dog or cat? People usually hire me when their pet is very old and has a few months left. They realize they only have iPhone photos and never invested in professional shots. Only when the animal is ready to go, do they realize what they will miss.

What do you hope to achieve? I want people to stop seeing animals as 'just pets' or 'just animals'. This simplistic view doesn't honour the important role of these creatures in our lives and in our ecosystems. Using photography, I hope to shine a different light on them, so they are valued, honoured and respected.

It is often said that 'photography can change the world'. Do you agree? Only we can change the world, but photography can help expand the right knowledge for people to take responsibility for their actions.

Are animal photos important? Photos, in general, are a *legacy*. A memoir of the times we live in. They convey our history and they tell stories. Stories of incredible times, once-in-a-lifetime events, people, places and animals.... They showcase animals that once lived and are now extinct. And they immortalize the ones that we know and love but one day won't be here. Photos are our time machine, to educate, to learn and to understand. And this is powerful.

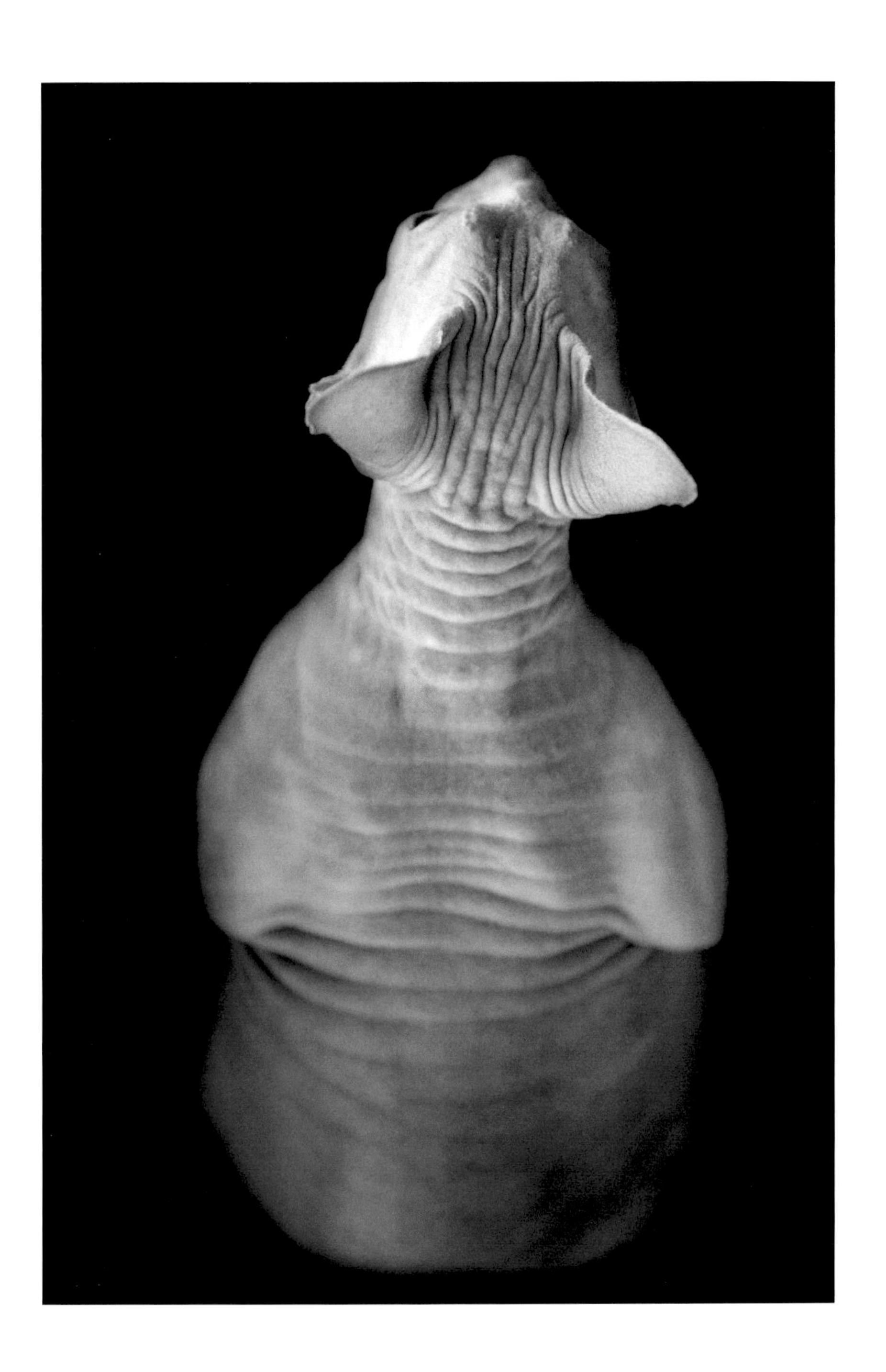

Jedi, a hairless Sphynx cat, San Francisco, 2013

Bear, an English bulldog, Los Angeles, 2018

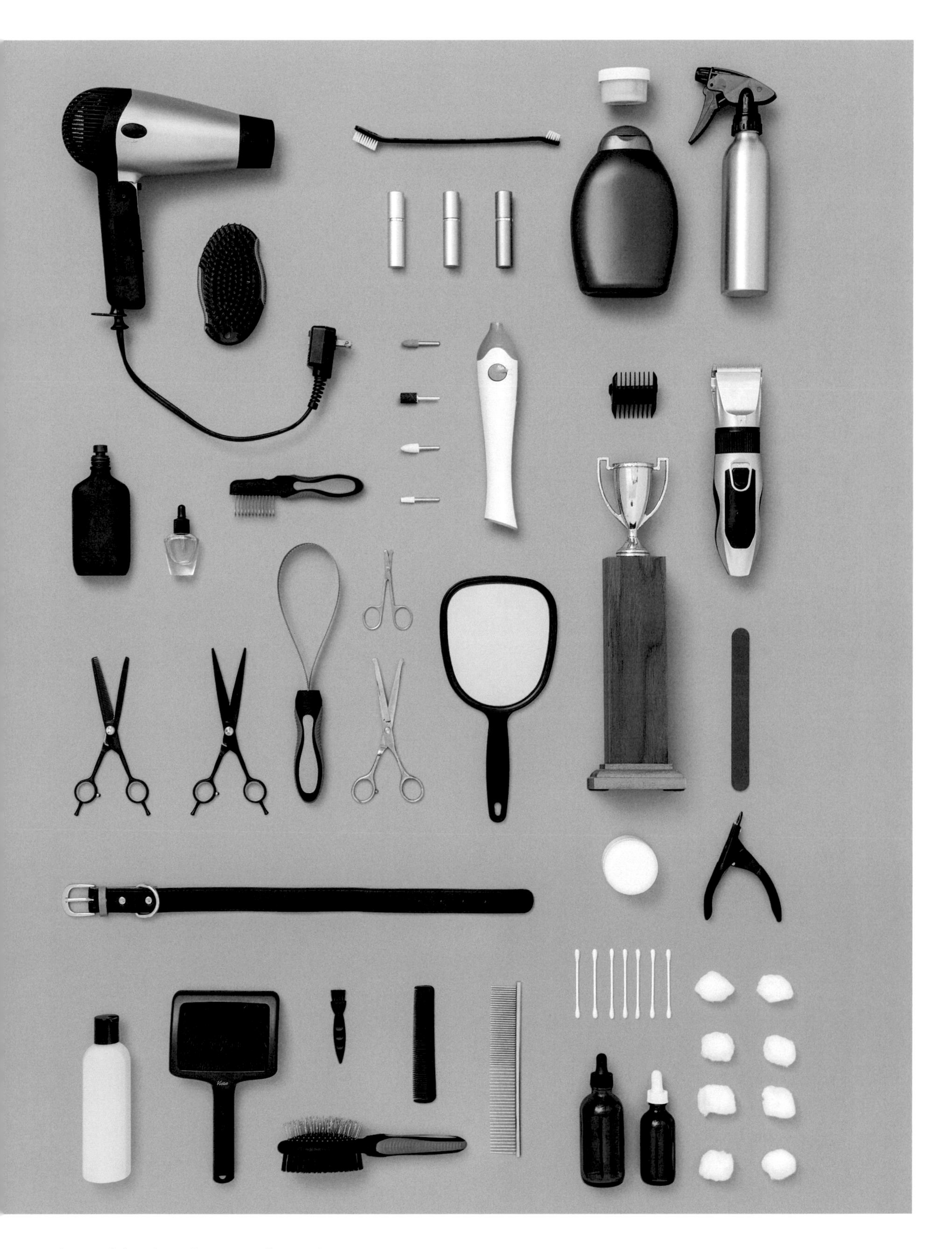

Zig, an Afghan hound, Los Angeles, 2018

Levon Biss

Natural Treasure

Levon Biss *is an award-winning British photographer. His passion for nature and photography came together in his celebrated* Microsculpture *project, a unique study of insects in mind-blowing magnification that took the genre of macro photography to a new level.*

I PHOTOGRAPH THE UNSEEN beauty of the natural world. That's the easiest way to describe what really motivates me. So far that means insects, fruits and seeds. Of course, photographers carry all kinds of intentions when they pursue their subject, sometimes these are things unspoken, or not fully realized, until years later when they look back but, for me, my *why* is simple. I create to encourage a love for all kinds of life – in particular insect life. I have a mantra that I repeat to myself often: make the invisible visible.

Photographs don't have to be grim or have uncomfortable subject matter to be important though. They can be challenging technically and that is a motivation for me. I'm the kind of person who likes to experiment and find a solution for how to photograph something in ways not yet done. It can be addictive. I'm working on bees for a new project. All very beautiful, of course, but a total killer to photograph under magnification. So many little hairs to get in focus!

I'm also continuing with another of my personal projects: insects caught in amber. I've been slowly searching them out and making my own collection. I have an expert dealer in Latvia helping me and when we get these precious artefacts back into the studio, it takes a while to mill the surface down and polish out the microscopic scratches to be ready for photography. You've got to shoot the amber in a liquid with a similar refractive index, so when light goes in you have some control and it doesn't just bounce around all over the place. I have an ant specimen that is thought to be up to forty-five million years old. Incredible.

* * *

My son Sebastian inspired me to begin the work for which I'm now well known. I was pretty exhausted back then, still working flat out shooting commercial work, sport and portraits mostly. My son, who was then six, came in from the back garden with a little black common ground beetle. We looked at the insect under a microscope and I was stunned by its beauty. I decided to photograph it for him so that he could be proud of making the effort to discover it. It was a challenge to work out the best camera set-up. In some ways it was like shooting a celebrity portrait – I'd take time with the lighting and work out the best way to make the subject look its best – but much less stressful than a celebrity!

Once I'd got to grips with bugs from the garden, I wrote to the Oxford University Museum of Natural History. They gave me access to their collections and that's when the *Microsculpture* project really took off. I was seeing insects under the microscope that I'd never even heard of – the orchid cuckoo bee, the splendid-necked dung beetle, the branch-backed treehopper – but they had interesting stories to tell.

The technique is still pretty laborious, but I'm perfecting it over the years. The lighting

Tricoloured jewel beetle, Oxford University Museum of Natural History, 2014

'Photography is a champion communication tool. If used well. It can be the most precise, the most harrowing, the most uplifting medium. It's the most direct form of communication we have.'

principles are much the same as for photographing a human. It's just a much smaller scale: you break it down; you treat each section like a still life; you forget about the rest of it; you just make that one area of the insect look as beautiful as possible; then you move on to the next section, which could be an antenna or the end of a leg. All the characteristics change and you re-light it and start again. Like a jigsaw, you hope that you have all the pieces when it all comes back together.

With photomicroscopy like this, the depth of field is so shallow that there's only a minute plane of focus. For each section of the insect to be in focus, it has to be photographed about six hundred times, with each photo taken about seven microns apart. That's about one tenth the width of a human hair! Those six hundred or so images are compiled into a digital composite. Each section is stacked and stitched, allowing for divergences in colour and perspective. It's pretty time-consuming when that is just one of fifteen or twenty sections of the jigsaw that make up the final image. A final insect portrait is the result of around ten thousand individual photos and takes four weeks to produce. That's a big commitment, but absolutely worth it.

I once gave a keynote speech at the annual conference of a microscope manufacturer, to a room full of scientists and experts. They liked the photos, but when I showed them the rig that I was shooting them on there were a few laughs. It was made of cable ties and timber two-by-fours. At the start I even had flip-flops built into the rig to try and absorb vibrations! The good folks at the company kindly sent me to Germany to pick out a very expensive microscope and an engineer came to help install it in my studio. But I had to send it back as the results couldn't beat that of my home-made set-up.

Many of the insects I photograph are barely more than a centimetre long, and the finer details of their surfaces could never be visible to the naked eye. That's what I want to reveal: their hidden beauty; things that we can't normally see. The entomological term for this texture is 'microsculpture', so that's how the project got its name. But I also want to photograph the animal in its totality, not just a pattern or a detail. I want people to be able to see the animal and begin a journey to learn more about it. Some people say the insects end up looking like jewels. That's interesting. The really vivid colours we see on some insects such as beetles are not due to pigments, but refraction of light off their delicate textured surfaces. One of the bugs in that collection, a tricoloured jewel beetle, was collected by Alfred Russel Wallace, who was a contemporary of Darwin. The specimen is 160 years old and still full of intense colour.

One particular moment on this project stands out for me: I was entrusted with a beetle specimen that Charles Darwin himself had collected in Australia and brought back to England on HMS *Beagle*. It was wonderful to feel close to history like that, working with something that had been found and treasured by such a significant figure. It adds another layer of interest to an animal that is intrinsically interesting even before we add a human story. That's the heart of it for me: creating photos to bring the stories of animal species to life in artistic and accessible ways.

Splendid-necked dung beetle, Oxford University Museum of Natural History, 2015

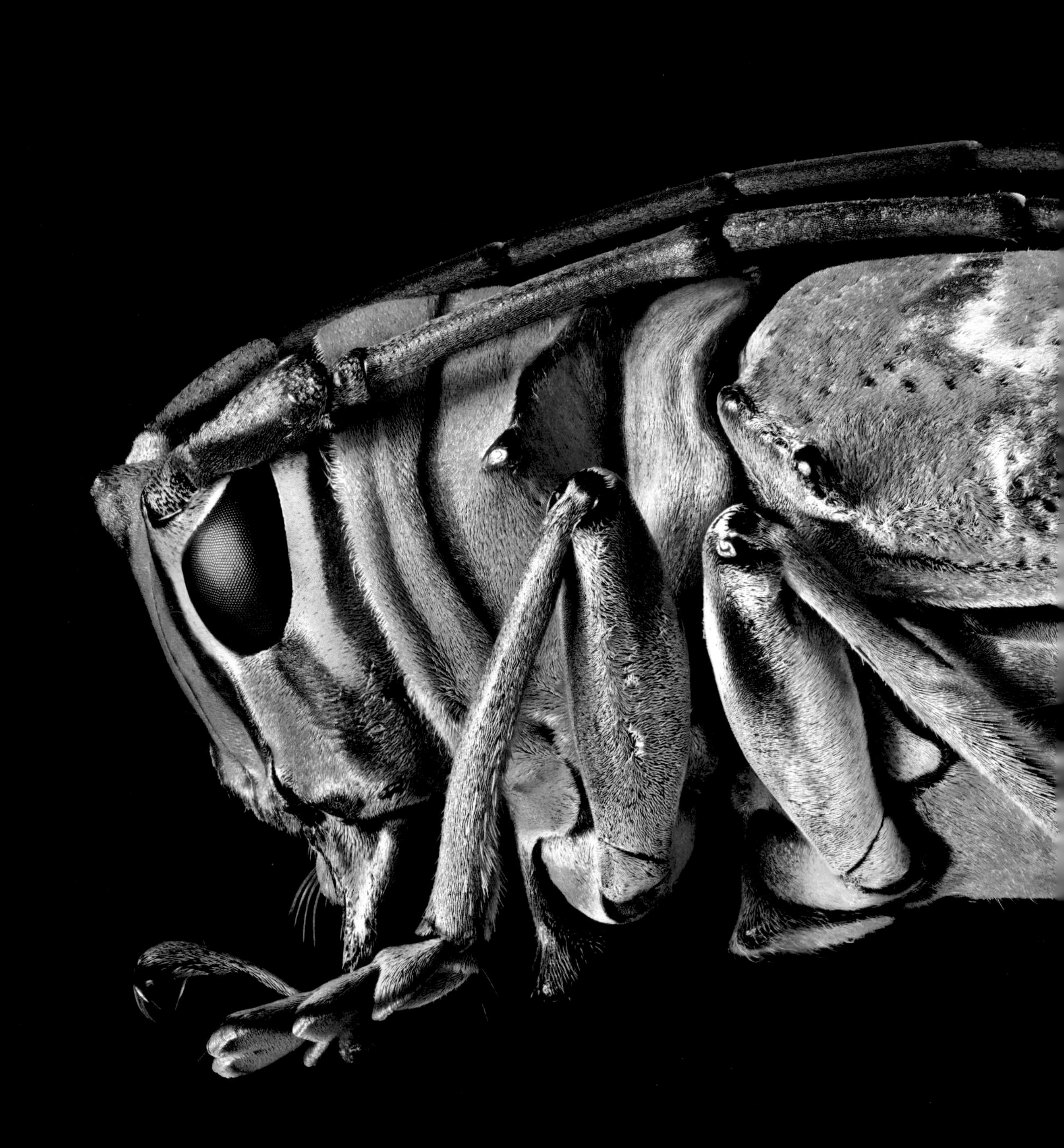

It is often said that 'photography can 'change the world', but I don't agree in the slightest. I certainly think photographs can inspire change for a brief moment, but if that moment isn't backed up with proactive governance then the effectiveness of the photograph and movement it creates dissipates pretty quickly. The idea that photographs can change the world is a romantic view. The world doesn't function on romance. I think we all know that.

But animal photos *are* important. There's a battle going on, with technological advances and easily digestible entertainment pouring into our brains. We're constantly on phones and we forget to *see* nature, to appreciate it. We're fighting against a rising tide of objects and data: a digital tsunami. We need reminders that there are other wonders out there, beyond our phones, our screens, our homes, other marvels that matter and are worth creating art and writing stories for. That's our challenge as image creators who focus on the natural world: to find a way to engage people and bring them back from the digital brink.

I want my work to be used as an educational tool. It can't just be another pretty picture that people admire for a second and then walk away from. I find that soulless, a waste of a communication possibility. It has to force the viewer to engage with it somehow. Look closer, lean in. To learn and take something away. It might just be a small piece of new natural history knowledge, but it might also be something that inspires them to study all kinds of new animals for themselves. Maybe a single photo makes a child realize that these bugs are engineering marvels. These tiny little insects, they're very clever, they're not just pests, or scary, or irritating. They're incredibly complex and absolutely vital for healthy, functioning ecosystems.

There is one question I get asked all the time: how do I get the insects to say still so long for ten thousand pictures? People usually giggle a bit when that one is asked at a lecture. Of course, they're specimens: they're long dead. But I also take that as a compliment. In photographing insects the way I do, I feel I'm able to bring them alive in authentic portraits that help us see things that we've never seen before; to appreciate creatures we've never even thought about.

There are millions of bugs living near us every day. I hope my photos encourage people to take even more care to help protect their habitats. Millions and millions of miniature animals, the natural treasure of our beautiful planet – we stand to lose it all unless we pay attention.

(previous pages) Jewel longhorn beetle, Oxford University Museum of Natural History, 2014
(opposite) Dark-winged fungus gnat trapped in fir-tree sap 40 million years ago, photographed 2018

Steve Winter

Hollywood Cougar

STEVE WINTER *has been a photographer for* National Geographic *for over two decades. He specializes in wildlife, particularly big cats, and speaks globally on conservation issues. He has been named BBC Wildlife Photographer of the Year and BBC Wildlife Photojournalist of the Year.*

THERE'S NO SPECIES on the planet that hasn't been affected by humans – and usually they are affected in a negative way. We need to tell their story and find a way to have empathy for them, then it can come full circle to us: if we have empathy for animals, maybe we will have more empathy for each other.

When *National Geographic* editors picked their best pictures of the last decade, my shot of the cougar (opposite) was one of them. When I was working on this story I went to meet the scientist who captured and collared these cougars north of Sunset Boulevard, in the Santa Monica Mountains National Recreation Area. He didn't really want anything to do with me at first, photographers were just a nuisance to him. I asked him a few questions and his answers were all 'no, no, no'. Then I said, 'you know, to really tell the story of urban wildlife, wouldn't it be great to get a picture of a cougar with the Hollywood sign?' And he looked at me like I was crazy. In fact, he told me I was crazy. Eight months later, he texted: 'CALL ME NOW!!!!' He had a bobcat-survey trail-cam picture of a cougar in Griffith Park, which is not far from the Hollywood Bowl, so we knew the shot might be possible – if we had the skill and patience to create it.

I spent fifteen months working on the shot, waiting for the cougar, or 'P-22' as this tagged cat was called. It turned out that he used the trail at night a lot, because most days he hid out in the undergrowth on the edges of the Forest Lawn cemetery where nobody bugged him. When we finally got the shot and the picture was published, people went nuts. There was a great upswell of public support: the LA school district started talking about wildlife; the local mayor made 22 October 'P-22 Day'; and now they're spending millions of dollars building the largest wildlife overpass in the world so animals can cross Highway 101 safely. When you think of an image like this you can see the power of photography to influence things positively.

* * *

When I began I really knew nothing about photographing animals. If you want to be in wildlife photography, I think a good place to start is knowing how to photograph people. If you can look at the world in a photojournalistic way, you're going to be a better photographer no matter what. For my first proper animal picture, I remember being sick with nerves the night before. I was in Costa Rica with a pharmaceutical company that was trying to find new drugs in the rainforest. It was my first shoot in the jungle. I was asked if I wanted to come see the *arribada,* when all the sea turtles come up the beach at night to lay their eggs. I knew I wanted to capture the moment, but what about just as dawn was breaking? I was pre-visualizing the shot and thinking on my feet. Maybe there might be an *abuela,* an old grandmother, who's taken all night to dig her nest. It was an outside chance, but that's exactly what happened. I lay in camp all day thinking, 'think of the turtle as an object, it's going to be coming towards me...what would

Mountain lion 'P-22', Griffith Park, Los Angeles, 2013

(previous pages) Snow leopard, Hemis National Park, Ladakh, India, 2008
(above) Olive ridley sea turtle returning to the sea after laying her eggs, Playa Naranjo, Costa Rica, 1992

‘Photography is everywhere and to me it’s everything. It enables us to see the world, to appreciate other cultures and to expand our minds. Photography is being open to all possibilities.’

I do with a person? They’re going to be in shadow so I’ll need flash fill…’ and so on. And that was my first animal picture. That beautiful turtle returning to the water after laying its eggs.

When I got back my editor loved the shot and that’s really how I got into animals. The impulse was to try things that hadn’t been done before, to find stories that would make us think about things differently. I also quickly realized the value in talking to people, listening well and trying to help them. Photography has to be a two-way process. A scientist once told me he hated photographers: ‘They take up so much of your time and we get nothing out of it.’ He mentioned needing pictures to raise money for his project and I gave him some to use. It was a smart thing to do because a little while later he emailed me and gave me an idea about the resplendent quetzal. What a gift a good idea can be – it became my first *National Geographic* wildlife feature!

I’ve since made photostories about jaguars to try and reverse negative attitudes towards the animal. A beautiful picture alone is not enough. All the cowboys in Brazil were saying every dead cow is the fault of a jaguar. I wanted to encourage people to see these big cats in a new light and to draw attention to the facts as they were being researched on the ground. The story we made supported the scientific studies being done on ranches there, and in the end they found that only about one per cent of cattle deaths could be blamed on jaguars.

It’s the same with snow leopards, which are often persecuted for losses of livestock in northern India. We made a story and raised money for livestock to be vaccinated – illness being the major reason for animal deaths. We started a project to vaccinate all those village herds, as long as the villagers promised not to kill snow leopards. So we’re making photographs of animals, but helping whole communities in the process. Ultimately with snow leopards, they’re best if we can just leave them the hell alone and respect the remote places where they live; but with a cougar in the Hollywood hills, we’ve encroached into their habitats so much now we really have to get involved, actively intervene to try and find a better way for animals and humans to live alongside each other. I tell the stories as best I can.

It’s worth adding, I rarely crop my pictures. The Hollywood cougar is exactly the way it came: no zooming in, no working, it’s just picture perfect. The flash didn’t bother him one bit, he just strolled on past, right through ten frames. If you look at the original RAW file, you wouldn’t believe how similar it is. And you better hope your RAW file is right, or you’re not going to win any Wildlife Photographer of the Year prizes: it’s important, as you want your pictures to have *life*.

I won that prize because my partner entered the contest for me. Because to me, I’d already won – I’d wanted to be a *National Geographic* photographer since I was a boy. I was living my dream. But I also understood that the exhibition travels all over the world, that so many more people have an opportunity to see those images, read the story and hopefully save that species. To me, it’s more important that our images help the animals that star in them. That’s the way your pictures can truly live.

SMART PATROL RANGER
PANTHERA

Tiger rangers study photos to identify individuals for conservation efforts and to combat wildlife crime, Thailand, 2010

Shannon Wild

SHANNON WILD *is an Australian-born wildlife photographer and cinematographer. She has authored several books, gives frequent photographic workshops and is an ambassador for a number of conservation charities and foundations.*

Describe an animal photo you've taken that was memorable to you. My inspiration, my pet – a central netted dragon, Raja. I remember taking a portrait of him and it conveyed his personality so well, he stood tall and confident with a cheeky look in his eye.

Do you have a favourite animal photo that you have taken? There are a few that come to mind but perhaps my favourite shows a white rhino with cattle egrets. Everything came together at that moment and to me it represents what it means to be a wildlife photographer: getting out and being in the right place at the right time and having the knowledge and experience to capture that moment. The cattle egrets were following this southern white rhino as it grazed, looking for bugs to eat that are disturbed by the rhino walking. At this moment several flew in towards the back of the rhino and I love how it looks like it could be one egret in different stages of flight, but it's actually three separate birds approaching. A beautiful moment topped off with a gorgeous moody sky as well.

What do you hope to achieve? In the moment, it's about doing justice to the subject, so I can celebrate its beauty and importance. I also enjoy the technicality of photography and filming. Ultimately, I want the finished product to hold meaning, to trigger emotion and curiosity. Sometimes that is for beauty's sake, to simply celebrate the subject, and sometimes it's to highlight the plight of a species, supporting scientific facts to make information more visceral. I believe both are important.

Has an animal photo forced you to examine your life? A starving mother orangutan and her baby. They had been stoned almost to death by villagers after being forced out of their habitat by fires lit to clear land for palm oil plantations. It is an image by International Animal Rescue that deeply moves me. I've been to Borneo a few times and seen the destruction first hand, and it made me reassess my product choices to avoid palm oil. I became an ambassador for Palm Oil Investigations, who create helpful product guides so consumers can make more educated choices, since the palm oil industry is devastating vital orangutan habitat.

Is there an animal photo that you most frequently think about? Peter Beard's aerial photographs of massive herds of elephants in Kenya from the 60s and 70s, in particular from his book *The End of the Game*, have really stayed with me. I would love to see that many elephants all together, but sadly due to various human activities, such as poaching and habitat loss, their numbers have greatly decreased since then. It would have been absolutely magic to witness.

It is often said that 'photography can change the world'. Do you agree? It has for me, so absolutely. It's a lofty ideal but, from personal experience, photographs opened my mind to the incredible wildlife of distant places and inspired me to seek them out. They've allowed me to express myself creatively, travel the world, earn a living and help bring awareness to endangered or misunderstood species. So, an emphatic 'yes'.

Southern white rhino with cattle egrets, South Africa, 2015

(above) Ground pangolin, South Africa, 2019
(opposite) White sifaka, Madagascar, 2015

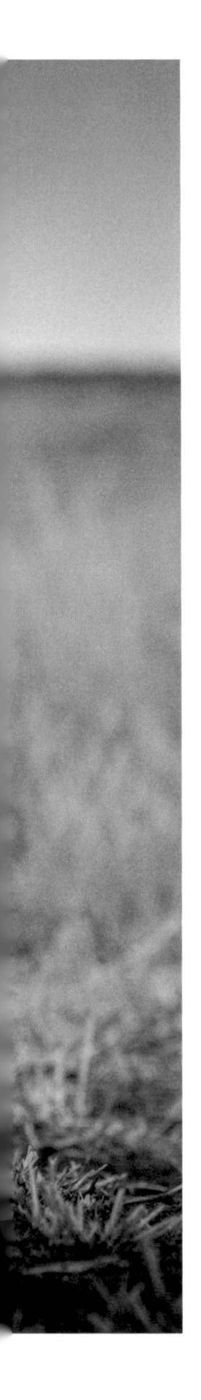

‘As a photographer I want to inspire compassion, whether it’s for a cute lion cub or a venomous snake.’

Shannon Wild

Black-legged kittiwakes, Kongsfjorden, Svalbard, 2017

Kiliii Yuyan

KILIII YUYAN *illuminates the stories of the Arctic and human communities connected to the land. Informed by ancestry that is both Nanai/Hèzhé (East Asian Indigenous) and Chinese-American, he explores the human relationship to the natural world from different cultural perspectives. Yuyan is an award-winning contributor to* National Geographic *and other major publications.*

Can you remember the first photo of an animal that you took? Honestly, when I was really little, I just spent a lot of time sneaking up on cats. Sometimes I would take pictures of them unawares, or play gentle tricks with them, you know, with a cucumber and seeing their reactions. I was trying to get a photo before I got noticed.

Do you have a favourite animal photo that you have taken? My current favourite at this moment is an aerial image of musk oxen gathered on the side of a hill outside of Kotzebue, Alaska. For me, some of that comes from how hard it was to get. The temperature was -26 °C (-15 °F), but about -38 °C (-37 °F) with the wind chill, gusting 45 knots, but I took a chance when the wind dropped every now and then. It was intense. The musk oxen themselves weren't fazed at all. Weather like this is no bother to them.

Is there a particular photo, or body of work, that raised your curiosity like no other? Erika Larsen's work with the Sami really piqued my curiosity because I wanted to know how she was able to get so close to the reindeer and their herders. And the answer? She was committed. She gave time to her work. She lived with a herding family for almost four years. That's a big lesson. Working in remote communities you must build trust and, in time, enter meaningfully into the rhythm of the animals that these cultures live with. You disappear into it. That's something that we have to teach our photography students too. To be patient. To be observant. And to be respectful – to animals and humans alike.

Can you describe the positive impact that one of your photos has had? I have a photograph of a hunter facing off with a polar bear. Every time that goes up on social media people are immediately up in arms and make the wrong assumptions. I find that photo has had an enormous impact because people initially question the hunter and the culture, but they end up having their misperceptions corrected and walk away with a much better understanding of Arctic life. So, what's important is the context and an accurate caption. The hunter was not trying to kill the bear but to protect the children from it. The hunter was ensuring that the families and children stayed safe from the very real danger of polar bears hunting on the sea ice. As photographers working in Indigenous communities, we have a responsibility to craft the story with understanding. You must choose carefully what you share and how it shapes the overall story. Not every picture needs to be released into the wild. Social media is a decontextualized place, where few people take the time to read behind an image.

Are animal photos important? Ecosystems need animals of all types to function. When the first recordings of whales singing underwater came out, feelings for these incredible animals dramatically changed. So much, in fact, that one of the things I have to deal with now is that people love whales so much they can't fathom the idea of hunting them. Just as it was for those whale songs, it is the same for powerful animal photographs. They bring us to places and show us things that are intimate, with context, to tell us things that we need to know, or remember. Understanding that Indigenous peoples have kept the natural world intact to this point and understanding Indigenous perspectives, including through photography, is critical to maintaining healthy wildlife populations into the future.

Flora Aiken, of the Iñupiaq, gives thanks to the first bowhead whale of the spring season, from the series *People of the Whale*, 2017

(above) Caribou, Teshekpuk Lake, Alaska, 2019
(opposite) A village dog gnaws on walrus remains, Chukotka, Russia, 2019
(overleaf) Musk oxen by drone, Kotzebue, Alaska, 2021

‘Storytelling about animals we encounter and the places we journey to is a timeless human impulse. A photograph always reflects who we are and how our culture sees.’

Kiliii Yuyan

PROFILE

Daisy Gilardini

DAISY GILARDINI *is a conservation photographer who specializes in the polar regions, with a particular emphasis on Antarctic wildlife and North American bears. In over two decades of polar voyaging, she has joined more than eighty expeditions and skied the final degree to the North Pole.*

Is there a particular photo, or body of work, that raised your curiosity like no other? I'm really inspired by Ami Vitale's body of work with rehabilitation centres in Africa and Asia (rhinos, elephants and pandas). I love how she can capture the deep bonds between rescued wildlife and caregivers. She brings awareness to the drama these animals face while at the same time filling me with hope for the future.

Have you ever risked your life for a photo? When it comes to wildlife, I generally think that if fear or danger come into play it means something is wrong in your approach to the situation. I believe the best ethic for wildlife photography is to photograph the animals on their own terms. This means positioning yourself in their environment and letting them decide if they want to interact with you. When they accept you as part of the landscape, they will reveal their personalities. In order to get intimate portraits of wildlife you have to be patient and never force the situation.

It is often said that 'photography can change the world'. Do you agree? Absolutely! Photography is a universal language and is understood by everyone, regardless of colour, creed, nationality and culture. As conservation photographers, it is our duty to capture the beauty of places and species that are at risk and raise awareness through the universal power of the images we capture. While science provides the data necessary to explain issues and propose solutions, photography symbolizes these issues. Science is the brain, while photography is the heart. We need both to reach people's hearts *and* minds, in order to move them to action.

Is there an animal photo that first changed your world? Without a doubt: *Polar Dance* by Tom Mangelsen, which has two polar bears sparring; *Blue Iceberg* by Cherry Alexander; *Emperor Penguins* by Bruno Zhender; and Norbert Rosing's work with polar bears. He was one of the first to take pictures of mothers with cubs emerging from the maternity den. Pioneering. And difficult!

Is there an animal photo that you see as your stand-out shot so far? *Motherhood* is for sure one of my signature shots. I think every mother in the world can relate to it. But more recently, *The Sleeping Bear* is also one of my favourite shots. I photographed spirit bears, one of the rarest bears in the world, in the Great Bear Rainforest for five years trying to raise awareness (among many other esteemed colleagues and NGOs) on the terrible project to build a pipeline through its territory. A feature article on this subject illustrated with my images was published in *BBC Wildlife* magazine in September 2015. This image, among a portfolio story on this issue, won the Nature's Best Conservation Story Award and was exhibited at the Smithsonian in Washington DC. The pipeline was finally rejected by the government in 2016.

Are animal photos important? People need to connect with the subject in order to care for it. Scientific facts, long articles, graphs and charts aren't nearly as effective as images in reaching people's hearts and emotions. There is a reason why we say an image is worth a thousand words.

(above) Motherhood, polar bear with her two cubs, Hudson Bay, Canada, 2016
(overleaf) Shadows, king penguins, Volunteer Point, Falkland Islands, 2018

‘Photography is not just an art form. It’s one of the most powerful mediums of communication we have.’

Daisy Gilardini

(above) The Sleeping Bear, spirit or Kermode bear, Great Bear Rainforest, British Columbia, Canada, 2012
(opposite) The Jump, brown bear trying to catch salmon, Katmai National Park, Alaska, 2010

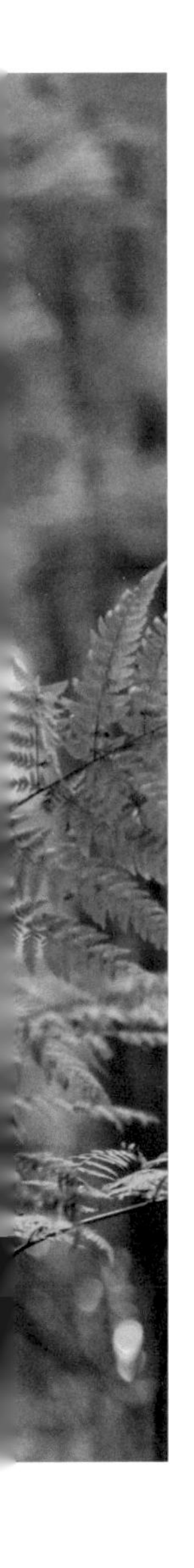

Anup Shah

ANUP SHAH *was born in Kenya, where he grew up exploring the national parks. He was featured in the book* The World's Top Photographers: Wildlife *(2004) and in* Hörzu *magazine as one of the five 'best wildlife photographers in the world'.*

Do you have a favourite animal photo that you have taken? The hippo, giraffe and gorilla photos featured here. The first two are black-and-white photographs taken with remote cameras, using a technique I pioneered almost twenty years ago. These are extreme wide-angle shots taken from ground level and they impart the six I's: Immediacy, Intimacy, Immersion, Inclusivity, Inside and Involved. The portrait of Malui, a western lowland gorilla, speaks volumes about her personality.

Have you ever risked your life for a photo? I've never been in a dangerous situation with animals. The welfare of the animals comes first and so I have to be extra careful to have them relaxed, doing what comes to them naturally. Suppose I put myself in danger to get a special shot and the animal got injured? The consequences would be that the animal would be shot by the national park authorities because it had become 'dangerous'. That is neither the truth, nor fair nor right.

So, why do you photograph animals? For a selfish reason: I have an urge for self-expression. What I want to say, my values, who I am, my philosophy – all are reflected in my photographs. I mainly photograph 'big game, in big country, under big skies' in East Africa, where I grew up. Through photography I feel connected to the land where our ancestors lived alongside the big game, perhaps in harmony. Certainly in greater connection than today. Maybe this has given me the urge to express that ancient relationship, something that I still feel.

It is often said that 'photography can change the world'. Do you agree? There are millions of things happening in the world at the same time as an image is published so it really is impossible to make too large a claim for photography. It's hard to disentangle the effect of one variable on an observed outcome. But all I can say from personal experience is that people do *feel* for animals when they see well-made, thought-provoking photographs – at least momentarily.

Is there an animal photo that first changed your world? There is a dramatic photograph of a leopard face to face with a baboon that was a double-page spread in *LIFE* magazine many, many years ago. It was created by John Dominis, who had been a war photographer in Korea. It's a very powerful image that has such intensity that you cannot get it out of your head. Many years later, John admitted that the photograph was set up, though originally shown as shot in the wild. I'd already embarked on the joyous career of wildlife photography, and I'm glad I did. But I make sure that my photographs are honest.

The shots in my *Serengeti Spy* book were all taken on the plains of East Africa. The trick in selecting locations lies in knowing the habits of the animals, that's all. Thus, in the dry season, I know of waterholes the thirsty animals will travel to; in the fruiting season, elephants will head for the fruiting trees I know of and so on. Of course, there is serendipity too, like finding a carcass. Unlike many animal photographers, I absolutely never try to attract an animal to my camera. However, my principle is simple and solid – animals come first, not the photograph. The effect of luring an animal, with bait for example, is to make the animal dependent on humans for food and to upset its natural habits.

Malui, a western lowland gorilla walks through a cloud of butterflies,
Dzanga-Sanga National Park, Central African Republic, 2011

(above) A hippo erupts from the water, Maasai Mara, Kenya, 2013
(right) A giraffe runs away from a lurking lioness, Maasai Mara, Kenya, 2012

‘We are putting more and more obstacles between our seeing and the real nature in the real world. If photography has one noble purpose, then it must be to see.’

Anup Shah

Vultures gather around a zebra carcass, Maasai Mara, Kenya, 2010

PROFILE

Will Burrard-Lucas

WILL BURRARD-LUCAS is a photographer, author and entrepreneur. He often seeks to capture close-up, intimate perspectives using a wide-angle lens. In 2009 he created BeetleCam, a remote-control camera buggy, which he took to Africa and continues to employ in his work today. He dedicates much of his time to working on long-term projects in Africa.

Can you remember the first photo of an animal that you took? A black bear on the coast of British Columbia in August 2001. I know this is exactly the moment I fell in love with wildlife photography. In that one hundredth of a second, my life changed. All of my passions came together: nature, technology, creativity and adventure.

Who are the wildlife photographers that you most respect? [Michael] 'Nick' Nichols, Steve Winter, Charlie Hamilton James, Ami Vitale. For many different reasons, but they also have a great deal in common: their artistry, the technology they employ and initiative they demonstrate, their ability to tell stories, the impact their work has had, their dedication to their projects and the sacrifices they make to get the shot.

What is the most influential animal photo ever taken? I think of vintage shots of the thylacine, known as the Tasmanian tiger. An amazing animal that humans persecuted to extinction and now only exists in a handful of photos and rare footage from the 1930s showing it pacing inside a cage. Might the influence of looking at ghosts like this, in old images, be to insist that we never forget? That we wake up and pledge to not let this kind of loss define us.

Is there an animal photo that you see as your stand-out shot so far? The black leopard. A photo of a black leopard under the stars was both the rarest and most difficult wildlife photo I had attempted. At any one time, I had between five and seven camera traps deployed. For the image to work, many factors needed to come together. Firstly, of course, it needed to be a clear night. But as the rainy season took hold, it became increasingly common for clouds to blot out the sky completely. It was also important that the night was very dark, with no moon or twilight to lighten the sky. To reveal stars, an exposure time of at least fifteen seconds was necessary. Several times the other conditions were perfect, but the leopard passed by in the wrong direction. Other animals were also making things tricky. I set up one of my cameras on a rocky outcrop on Suyian Ranch with spectacular views out over Laikipia, a scene that showed the essence of this land of leopards. Within a few days the camera captured a photo of a young leopard at dusk, but shortly thereafter, a troop of baboons discovered the camera and comprehensively tore it apart. Eventually, after six months of perseverance, I returned to a cluster of three cameras I had set up on a promising rock. One of the cameras was an infrared DSLR, which I had just set up to capture a behind-the-scenes picture. I eagerly checked the remaining cameras and, on the lower of the two cameras, I had the images of my dreams!

What does photography mean to you? Photography is about preserving moments, capturing beauty, evoking emotions and touching people. A big part is not just about the photograph itself, but sharing it with the world so that it has the opportunity to impact people.

Lion cubs, Maasai Mara, Kenya, 2011

(above) Black leopard, Laikipia, Kenya, 2019

(overleaf) African wild dogs, South Luangwa National Park, Zambia, 2012

Paul Souders

On My Own

PAUL SOUDERS *is a wildlife photographer whose work has sent him across all seven continents. In 2013, he was awarded the Grand Prize in the National Geographic Photo Contest.*

YOU COULD SAY that it's all my grandmother's fault. Irma was an avid amateur photographer who regularly travelled on international package tours back in the 1960s, part of that first wave of American tourists who set off to explore the world after the war. She visited Paris and Cairo and Hong Kong, went up the Ganges and down the Amazon. All the time carrying her clunky old Nikkormat SLR on a weathered leather strap around her neck. After she got home, the family would all sit in a darkened room and watch her slideshows. It seemed like another world from our little house in rural Pennsylvania.

She gave me my first camera, a Kodak Instamatic, on Christmas Day in 1974. It shot 126 cartridge film. I raced outside to make twelve frames of bare winter trees in grey light, then waited in agony for a week before my little masterpieces came back from the lab. I do dimly remember the pictures were a disappointment, but I was not easily discouraged. If my grandmother could make great shots and travel widely, what was going to stop me? Other than talent, money and opportunity of course!

That little camera hung off my wrist for years. I spent whatever allowance money I could scrounge on film, then went out and shot pictures of my Little League baseball games and model rocket launches, my dad's hunting trips for local deer, or family vacations to the seashore. It's a shame those pictures are all gone. Not for any artistic merit, but as a snapshot of what seems now to be a very distant time. Not counting the adopted mutt I grew up with, I guess the first animal I photographed was a white-tailed deer wandering through our backyard before the hunting season started.

I never set out to be a nature photographer. Where I'm from nature is mostly junk cars, poison ivy and rusted barbed-wire fences. I hoped to grow up to be a famous war photographer. I was obsessed with the old heroes like Robert Capa and W. Eugene Smith when I first started out. I found a copy of *Let Truth be the Prejudice* at our local college library, such beauty and truth in these images of human suffering. But I think Galen Rowell's *Mountain Light* wound up steering me towards trying to capture the wonders of our natural environment.

I spent the first decade of my career working as a newspaper photojournalist. Even at the small papers where I started out, I saw enough human misfortune and pain to recognize that I would be well served to find another line of work. I stumbled into a gig at a newspaper in Anchorage, Alaska. All of a sudden, I could wake up to a moose on my front porch, bald eagles on the drive to work. I grew up between a chicken farm and a trailer park, so this was all new to me. In between regular assignments, I began to explore the wilderness that surrounded the town. The landscapes and critters I saw were a lot more interesting than most of the stuff I was getting paid to photograph.

Still, I was pretty sure I'd die of old age before my bosses at the paper, let alone the editors at *National Geographic*, got around to paying me to shoot the assignments I dreamed of. I knew it would be easier to spend what little

Polar bear, near Sentry Island, Hudson Bay, Nunavut, 2013

‘Photography is the struggle to find magic and wonder in the everyday world, then somehow capture it within the borders of a frame, like lightning in a bottle.’

money I had and just go. So I left that steady paycheck behind, packed up the truck and set out. Living hand to mouth, I made just enough money each month from photo sales to keep myself in ramen noodles and cheap beer – and keep moving. Along the way, from Alaska to Southern Africa and the Antarctic, I discovered a whole new world of large and dangerous animals.

* * *

I had been making noise for years about heading up to Hudson Bay to see polar bears. Not with all the tourists, squashed into a Tundra Buggy, but out on my own. I landed on the idea of walking out my front door and setting off alone to find them. I decided this was going to be a true BYOB job: bring your own boat. I loaded my truck with a ten-foot inflatable, outboard motor and pretty much every stitch of warm clothing and working camera gear I owned, then set off across 1,500 miles of highway to reach the railroad terminus that led north through the boreal forest to Churchill, Manitoba.

When I decamped some twenty hours later in Canada, the calendar said it was late June, but I was greeted with the hard slap of rain whipping in off the pack ice. It was just me and that mountain of gear on the platform as the train pulled away. The truth was that I had no idea what I was doing. Small boat, big water and aggrieved polar bears; it sounded like the perfect recipe for grisly headlines. But as soon as the wind dropped, I ventured out onto the water, nervously picking my way through the ice, scouring the horizon for some hint of a polar bear. I spent hours staring at the floes over the days and weeks that followed, motoring hundreds of miles through the melting pack. As I wound my way through the shifting maze, I stopped frequently, climbed up onto any high spot I could find on the ice, then slowly scanned the horizon with binoculars for the white-on-white outline of a bear. I stayed out until the midnight sun dipped below the horizon, leaving me to navigate home by GPS and the distant twinkling lights of town.

In mid-July a line of squalls roared through and summer arrived in the form of record heat, biting black flies and a strange haze – smoke from distant forest fires – filling the sky, drifting in from the south. In the orange half-light of a setting sun, every lump and hummock looked like a bear. Hours passed. From a crumbled snow-covered ridge, I looked and then looked again and – to my astonishment – saw movement. Half a mile away, a young bear woke and quickly shambled from the ice towards the water. Sliding ass-first off my own iceberg, I hopped into my boat and set off, struggling to keep the bear in sight.

Most bears will give human contact a wide berth, but this one, a young female judging by her size and build, quickly calmed and began to grow curious as I slowly trailed her. We were soon moving through the water in tandem, separated by a hundred yards, then fifty, then – holy shit, that bear was really close.

I dumped my camera gear out of its waterproof cases and shot her with the works: telephoto lens, wide-angle lens, underwater pictures with a housing and fisheye lens. I held the outboard’s throttle and steered the boat with one hand while shooting with the other. I even mounted one camera onto a six-foot boom and then awkwardly tried to swing it closer to her.

(top and above) Polar bear, Hudson Bay, Nunavut, 2013
(overleaf) A polar bear hides submerged beneath melting sea ice, Hudson Bay, Nunavut, 2012

I succeeded only in dunking the contraption into salt water, killing the camera, lens and trigger. Undeterred, I dug out a spare camera and began chewing the insulation off a copper wire to jury-rig a replacement shutter cord. As the bear swam beneath an iceberg, I managed to drift the boat in closer and hang the boom beside a hole in the ice. She rose to breathe and I began shooting, blindly pressing the shutter cable and hoping that something, anything, might be in focus. She submerged for a moment, then surfaced again for one more breath before disappearing beneath the ice.

The midnight sun hung like a dying star in the hazy orange sky. The bear reappeared and paddled slowly towards the sunset on a sea glowing like molten metal. The moment felt like I had been given a perfect jewel, something precious to hold on to for the rest of my days. I could have followed that bear for hours through the short half-lit summer dusk. But I cut the engine and let the boat drift. I watched in silence as she swam away, a slow vanishing.

On the southbound train home a week later, I sorted through the photographs on my laptop. There was my bear, walking across the ice, swimming and diving. Suddenly, there it was: one magical image that I'd never seen before, nor imagined, not even in dreams. In it, the polar bear floats beneath the surface, staring back up at my camera, surrounded by ice and empty sea, lit by the burnished, hazy sun. I laughed out loud, then started parading up and down through the passenger cars like some lunatic, showing the picture to a trainload of complete strangers. In forty years of working as a professional photographer, that's the best picture I've ever made.

Polar bear, Harbour Islands, Nunavut, 2014

Fragments

Martin Parr, seagulls eating chips, from the series *West Bay*, 1996

Eric Hosking, a barn owl returns to its nest, Suffolk, 1948

Unknown, 'National Collection of Heads and Horns', Bronx Zoo, New York, 1910

Richard Barnes, man with buffalo, from the series *Animal Logic*, 2007

David Chancellor, trophy room, from the series *Safari Club*, 2013

Laton Alton Huffman, 'His First Grizzly', stereograph, 1882

Werner Bischof, boys admiring photos of the bullfights, Mexico City, 1954

lliott Erwitt, chihuahua contact sheet, New York City, 1946

Matt Stuart, *Trafalgar Square*, London, 2007

Richard Peters, *Shadow Walker*, Surrey, 2015

Henri Cartier-Bresson, Henri Matisse sketching a dove, Vence, France, 1944

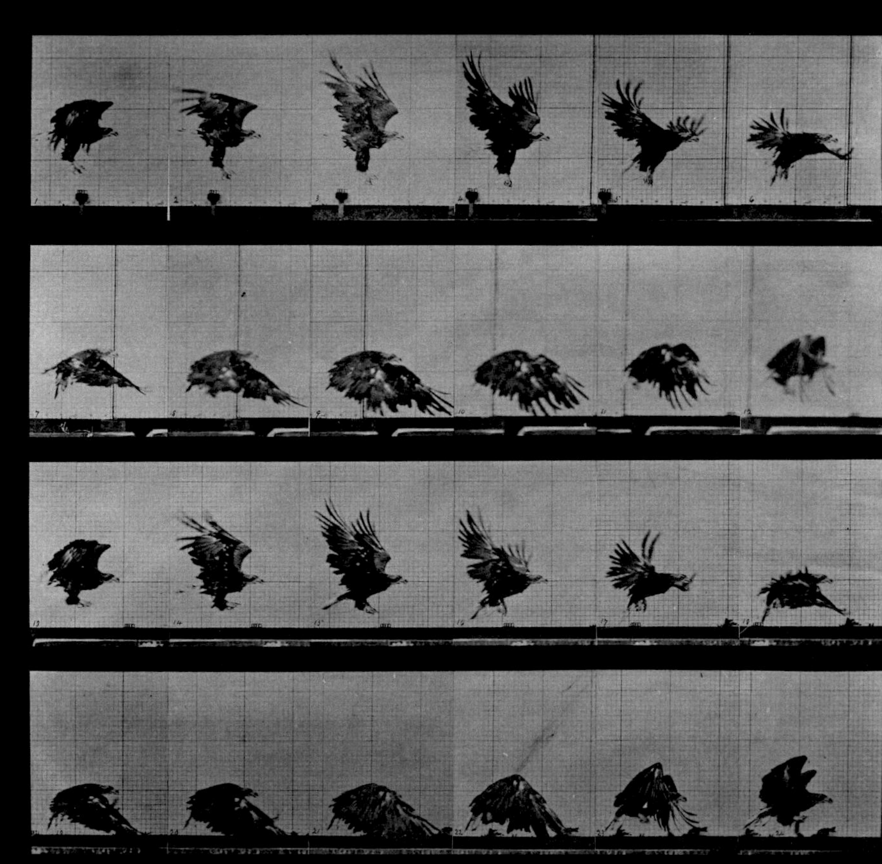

Eadweard Muybridge, vulture in flight, from the series *Animal Locomotion*, 188

Stephen Gill, starling flypast, from the series *The Pillar*, 2019

Dmitry Kokh, polar bears at an abandoned meteorological station on Kolyuchin Island, Chukotka, 2021

HISTORIES

Creative Doing

‘I do not profess to have perfected an art, but to have commenced one; the limits of which it is not possible at present exactly to ascertain.’

William Henry Fox Talbot

THE EARTH IS 4.6 BILLION years old. If we scale that to forty-six years, to try and get a sense of this vast amount of time, humans have been here for just the last four hours. The Industrial Revolution began one minute ago. In that time, we have destroyed more than 50 per cent of the world’s forests. This is clearly not sustainable. Right now, we use 100 million barrels of oil every 24 hours. Each day we lose about 200 species, a rate of loss that places us firmly in the midst of a sixth mass extinction event. We spend billions of dollars searching for life on other planets and trillions of dollars killing the life on this one. When will this madness stop?

We need to rethink and reimagine a future for ourselves and our animal kin. Statistics may not push us to take climate action but stories can. ‘Tell me the facts, and I’ll learn’, runs a Native American proverb. ‘Tell me the truth, and I’ll believe. But tell me a story, and it will live in my heart forever.’ Or, as Richard Powers put it: ‘the best arguments in the world won’t change a person’s mind. The only thing that can do that is a good story.’

Stories shape our lives: how we live them and what it means to be human. ‘Artists must confront the climate crisis’, says Ben Okri, ‘we must write as if these are the last days’. Storytelling has real power, in our brains and in our hearts. It is through stories that we dare to imagine a better world. Creativity can drive awareness and shift perspectives, and the stakes could not be greater. In words, in images, in actions. Making artworks, joining marches. Creating new narratives. Finding new ways of doing and being.

In the long history of human–animal relations, photography is a very recent innovation. And yet, it has dramatically changed the way we understand and value animals. As John Berger has written: ‘zoos, realistic animal toys and the widespread commercial diffusion of animal imagery, all began as animals started to be withdrawn from daily life. One could suppose that such innovations were compensatory. Yet in reality the innovations themselves belonged to the same remorseless movement as was dispersing the animals’.

The *how* we photograph has grown in lots of different directions; the *why* we photograph, it might be said, has also evolved. Many of the photographers interviewed in this book are tireless in their campaigning

William Henry Fox Talbot, wings of a lantern fly, 1839
–
Talbot experimented creating images with a solar microscope. ‘After all, what is Nature,’ he wrote, ‘but one great field of wonders past our comprehension?’

for conservation causes and they photograph to help protect the animals they care about. Of course, in the earliest days of photography this kind of motivation to make images did not yet exist, but a passion to experiment with new techniques, even an obsession to create, certainly did.

We might never know for certain why Montizón made an image of a hippo, or why Leen was later so drawn to bats; why Shiras, night after night, paddled his canoe into the darkness, or why Kearton stood in a field, for hours on end, dressed as a tree – but it is safe to say that curiosity was an intention that unites them. And a little love, surely. We choose to spend our days photographing animals because there are few things we would rather be doing. Even though the thought of enduring blizzards or endless early mornings doesn't sound like much fun at all. I've always liked what Jean-Louis Etienne said when asked why he went on polar expeditions: 'Because I like it', he replied. 'You never ask a basketball player why he plays: it is because he enjoys it. It is like asking someone why they like chocolate.'

Creation can also be a cure for despair. For many photographers, working out in nature and with animals in particular helps them find purpose in this age of climate crisis. They make photographs to encourage environmental awareness and to help build a society that values the natural world, and you can too.

With admiration and with art, we raise our cameras as tools of advocacy and action. 'Generation Dread', as Britt Wray has described it, is overwhelmed as never before, and with a despairing diet of daily news and doom-scrolling that's no surprise. But it's also incredible to see how people are rising up. Not being willing to accept the status quo or, far worse, to accept the fate to which a lack of action on climate will condemn us. We must not hide from the hard reality of life on our crowded, injured, but still very beautiful planet. We have to move forward in positive and constructive ways.

Creativity is something that empowers and enables us to feel less climate anxious. Hope itself is a creative practice. Hope is not wishful thinking but being active and ready to engage. Chase Jarvis has it right: 'the best antidote to negative thinking is creative doing.'

* * *

When I'm not leading voyages in the Arctic, or trying to make books, I teach natural history and photography at a university in Cornwall. Our course has evolved from portfolios of pets and puffins, to thoughtful and ambitious projects, deeply felt and often international in scope.

Herbert Ponting, self-portrait posing with his telephoto camera, Antarctica, 1912

–

'Without my cameras I was helpless', Ponting wrote during that expedition. 'At all costs ... my precious kit should be saved. We would survive or sink together.'

'Instructions for living a life: Pay attention. Be astonished. Tell about it.'

Mary Oliver

(right) Marsel van Oosten, snow monkey, Jigokudani, Japan, 2014
–
'Most animals are curious about unfamiliar things', Marsel says. 'Elephants like to pick cameras up and throw them away.... Rhinos push them over and large carnivores like to bite them.'

(overleaf) Sergey Gorshkov meeting a curious red fox, Kamchatka, 2008
–
'This is my life now,' Sergey says, 'My favourite places are the Russian Arctic and the taiga forest. There is an age-old peace and silence everywhere. You stop counting, you join the days and the world of wild nature. Each of us has our own dream and plan for life, and each our own taiga, our favourite impassable corner.'

Just as I asked the professionals featured in this project, I regularly encourage my students to answer the simple but difficult question of *why* they photograph animals.

Watching all kinds of animals in the wild – observing, in joy and gratitude – can lead to personal fulfilment. It might be whales, or wagtails, or worms, in the woods, or in wild waters. Whatever it is, you should immerse yourself in what you love, filling your life with meaning. These are the simple things I wish for my students, as for everyone. That meaning might come from appreciating animals, or in committing yourself to understand more of the ecology of which they are a part. It might come from helping others, from learning something new or in sharing your skills. Educating yourself and those around you.

The answers my students give me to *why* they photograph animals gather around similar themes; words like fascination, admiration and curiosity come up regularly. They create images because they are interested and because they care, because they like to spend time in nature. They use photography to learn more about animals and to think more deeply about how we humans see the world. To play a part in conservation and science communication and to raise awareness of the issues that animals face. To be better global citizens. To be active.

So, why photograph animals? 'Because I enjoy being outside', says Ben, 'and want to show the beauty of the world around me.' 'Because it makes me happy', says Nicole. 'Because they're way cooler than people', says Lali. And says Bryony, 'because, *why not*?'

Chronology

LIKE THE EVOLUTION of species, the history of photography is not a simple story of a defining moment when everything comes together, exploding into life fully formed. There are major discoveries and technical innovations, of course, but more so very gradual changes, mistakes, intriguing new techniques and frustrating false starts. Progress comes in fragments. Delight, sure, but debts and delusion too. Success, yes, and fame for some, but also huge sacrifice: cracked plates, failed exposures, lost kit, broken dreams, wet days in the field and nothing to show for it. Yet for the really committed, curiosity always wins out. Photography comes into focus slowly and surely over many years, with men and women from many nations extending what an image might mean: pushing themselves and their cameras into new realms of possibility. From daguerreotype to digital, from the studio and into space, animals are an essential part of this remarkable story.

Frank Haes, cheetah at London Zoo, 1865

30,000 YEARS AGO
The earliest pictures are made with natural pigments and stone tools. Animals are present at the very beginnings of visual culture, carved in ivory and bone, and daubed and inscribed on cave walls.

1826
Nicéphore Niépce is experimenting with a process to capture an image of the world. The first photograph that survives shows the view from the window at his country estate in Saint-Loup-de-Varennes. We can just make out a pigeon loft, a pear tree, a slanted barn roof and the bakehouse chimney. Niépce calls his process heliography, from the Greek *hēlios*, meaning 'drawing with the sun'.

1838
Louis Daguerre makes the first photograph of an animal: a human, having his shoes polished, standing on the corner of the Boulevard du Temple. The exposure time for the image is around seven minutes and as a result any moving things – people, carriages, horses, dogs – disappear from the scene.

1839
The invention of photography is announced to the world. William Henry Fox Talbot reveals his 'photogenic drawings'. Among hundreds of experimental shots, he has made intricate images of insect wings using the focusing power of a solar microscope. Daguerre releases details of his methods to the French Academy of Sciences and within months his technique is being used across the Atlantic.

Louis-Auguste Bisson, a Cleveland Bay Stallion, 1841

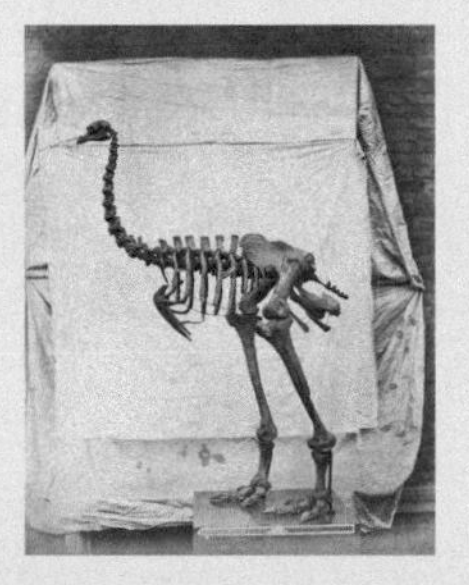

Roger Fenton, skeleton of the extinct moa, 1854

John Dillwyn Llewelyn, boy with lion cubs, 1854

Harry Pointer, self-portrait with his pets, 1870

1840

At home in Bristol, artist Sarah Anne Bright is experimenting placing leaves and other objects onto sensitized paper, inspired by Talbot's technique. Of the few images that survive in her brother's album: a leaf and the egg case of a catshark. Alfred Donné, a French doctor, creates detailed images of the eye of a fly.

1841

Louis-Auguste Bisson learns the daguerreotype process directly from Daguerre and begins photographing professionally. He captures an image of a dog for the animal artist Rosa Bonheur and is encouraged by her to picture horses. His interests shift from daguerreotype to photography on paper, and his skills take him all over Europe.

1842

Joseph-Philibert Girault de Prangey captures one of the earliest living animals rendered in a photograph: a camel in the desert near Alexandria.

1843

Anna Atkins is making cyanotypes of seaweed. She gathers them together as an edition, *British Algae*, the first book produced entirely by photographic means. The following year, the first instalments of Talbot's *The Pencil of Nature* bring photographic images to the attention of a wider public.

1848

Edmond Becquerel, a French physicist studying the solar spectrum, experiments with metal plates sensitized with silver chloride. He makes images of stuffed birds and coloured engravings – some of the first successful reproductions in colour. A plate depicting a parrot shown at the 1855 International Exhibition in Paris was particularly admired.

1852

Don Juan the Count of Montizón photographs captive animals in London Zoo. His shot of hippo Obaysch is entered into an album given to Queen Victoria. Some of his photos are displayed in 1854, one of the earliest photographic exhibitions.

1853

Bisson creates illustrations for *Photographie Zoologique*, a nature encyclopaedia and the first scientific publication illustrated with photographs.

1854

Roger Fenton makes images of collections in the British Museum as its first official photographer. His shots include now extinct animals like the megatherium, moa and Irish elk. William Bambridge is appointed Royal Photographer to Queen Victoria. Beyond creating endless portraits of the family, he creates many images of their pets, as well as Prince Albert's hunting in Windsor park.

1856

John Dillwyn Llewelyn creates artful shots of deer, heron and otter on his estate in Wales. Exposure times are still too slow to capture wildlife with clarity: he uses taxidermy specimens instead. William Thompson attempts to shoot photographs underwater in Dorset, building a wooden box to house his camera attached to an iron tripod and lowering it over the side of his rowing boat. Unfortunately nothing of use survives.

1858

James Chapman takes wet-plate camera gear to Africa to photograph elephant and zebra in their natural habitats, but he returns only with photos of dead animals shot on the expedition.

1860

James Black creates images from a hot-air balloon above Boston. One good print results from the eight plates he makes, which he titles *Boston, as the Eagle and Wild Goose See It*. It is the oldest surviving aerial photograph.

1862

Allan Scott is photographing in India. His 'stereogram' images of Hyderabad and its peoples are among the earliest known, and also include a dead tiger and rare Asiatic cheetah.

1863

Léon Crémière photographs dogs at shows in the Bois de Boulogne, chaining them to trees or against a sun-lit wall in the hopes of getting a good shot. To avoid the blur caused by a wagging tail, he sometimes ties them with string.

1865

Frank Haes photographs captive animals in London Zoo. He makes and sells prints, early stereoscopic images. Over the decades that follow, many photographers meet exotic animals for the first time here, like Thomas Dixon, Frederick Bond and J. E. Saunders, who even shoots a series on Kodak for cigarette cards.

1866

Harry Pointer begins photographing his pet cats in funny poses in Brighton. By March 1872 there were more than a hundred to choose from, and numbers keep multiplying.

1869

George Critcherson and John Dunmore, studio photographers from Boston, accompany the artist-explorer William Bradford on a voyage to Greenland. Their images of towering icebergs and polar bears are published in *The Arctic Regions* in 1873. Victor Prout is in Tasmania and photographs a thylacine, the

Louis Ducos du Hauron, rooster and parakeet, 1872

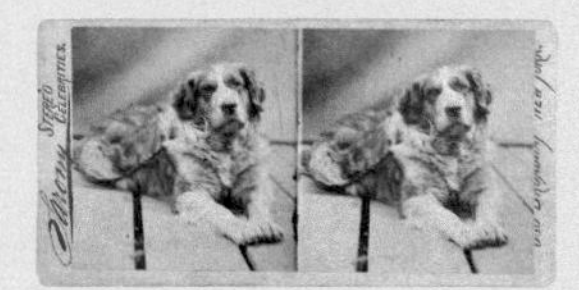

Napoleon Sarony, 'celebrity stereograph', 1880

Ottomar Anschütz's electrotachyscope, *Scientific American*, 1889

Josef Maria Eder and Eduard Valenta, X-ray of an Aesculapian snake, 1896

Richard and Cherry Kearton, *Wild Life at Home*, 1898

'Tasmanian Tiger', recently shot by Hobart chemist William Weaver.

1870

Frederick York's photographs at London Zoo include the last-known image of the quagga, Obaysch the famous hippo and Jumbo the African elephant, later owned by circus showman P. T. Barnum. Charles Hewins of Boston photographs a white stork on its nest in Strasbourg, often said to be the first image of a wild animal in its natural environment. Critcherson and Dunmore's polar bears are earlier but they don't survive the encounter.

1872

Louis Ducos du Hauron is trying out practical processes for colour photography. An amateur physicist and artist, he shoots all kinds of things at home: taxidermy birds, botanical specimens and views from his window. His three-colour layers when superimposed make a single image: a technique that is essentially the basis of modern colour processes.

1873

Caleb Newbold photographs rockhopper penguins on Inaccessible Island in the South Atlantic during the scientific cruise of HMS *Challenger*.

1876

Eadweard Muybridge creates a photo sequence of a galloping horse to settle a bet on whether all its hooves leave the ground at the same time. He later applies his technique to other animals including wild deer, which could be the first true wildlife to be captured in any kind of motion picture process.

1880

The craze for stereographs dominates photographic culture. They are paired images, printed and mounted on card, for viewing in a holder through a set of special lenses (stereoscopes), giving armchair travellers a three-dimensional view of the world.

1881

Étienne-Jules Marey, a Professor of Zoology at the Collège de France, is developing a 'photographic gun' capable of shooting rapidly sequenced photos from a single lens to better understand how animals move: his desire is 'to see the invisible and push the boundaries of our senses'.

1882

Ottomar Anschütz develops a machine (known as an electrotachyscope) to show a sequence of photos on glass plates and, later, images printed on celluloid or mounted on card. Sequences include dogs and horses, a flying stork, a leaping goat, a camel running and a man riding an elephant.

1884

Muybridge 'films' a staged encounter of a tiger killing a buffalo at Philadelphia Zoological Gardens in the first, but certainly not the last, instance of fakery in wildlife film-making.

1888

American businessman George Eastman invents photographic film. At last photographers are freed from the burden of heavy boxes of plates and chemicals, and his roll film is quick and easy to load. Eastman's 'Kodak' camera hits the market. Small, cheap and portable it proves an instant hit.

1890

Paul Nadar journeys to Central Asia and brings back several pictures taken with a Kodak. In Turkestan, the Emir of Bukhara arranges a falcon hunt in his honour. After decades of experimentation and partial success, a method of cheap but accurate photographic illustration is becoming widespread. The halftone process enables the reproduction of prints in books and magazines at an unprecedented scale.

1892

George Shiras begins his photographic explorations on Michigan's Whitefish Lake. Shooting from the front of his canoe, he uses a hunting technique called jacklighting to illuminate wildlife emerging from the woods at the shoreline. Louis Boutan, a French biologist fascinated by molluscs, begins taking photos underwater at Banyuls-sur-Mer. His equipment is so heavy it was said to take three men to move it.

1895

Max Skladanowsky films a boxing kangaroo and its circus trainer, arguably the first animal presented to the public in a modern motion picture. Richard and Cherry Kearton publish *British Bird's Nests*, the first wildlife book illustrated entirely with photographs. Louis Lumière films a lion pacing its cage at London Zoo as a keeper tosses meat through the bars.

1896

Just a month after physicist Wilhelm Conrad Röntgen's announcement of the discovery of X-rays, the Austrian photo-chemists Josef Maria Eder and Eduard Valenta publish an elegant volume of photogravures depicting the skeletal structure of animals and human body parts. The public are mesmerized by these hauntingly beautiful and revolutionary images.

1897

James White films sea lions off San Francisco. His footage is released as *The Sea Lions' Home*, perhaps the first true

George Eastman's Kodak camera, advertisement, 1900

Carl Georg Schillings, *Mit Blitzlicht und Büchse*, 1905

Emma Turner, nestling bittern, 1911

John Ernest Williamson going down to take photos, 1913

wildlife film, showing real animal behaviour in a natural environment. The Kearton brothers publish *With Nature and a Camera*, revealing their innovative techniques for photographing animals in their habitats. The following year, their *Wild Life at Home* is the first how-to guide for keen nature photographers. It's also notable as one of the first uses of the term 'wild life' in connection to visual images.

1900
On expedition in East Africa, Carl Georg Schillings takes the first flash photographs of wildlife. They appear in his book *Flashlights in the Jungle* (*Mit Blitzlicht und Büchse*) released in 1905 and are said to have influenced many to ditch their guns and take up 'camera hunting' instead. But his ethics are questionable: a tethered donkey is used as bait to lure lions. The arrival of the Kodak box 'Brownie' with its sensational price of just one dollar is a worldwide success.

1901
Allen and Augusta Wallihan are the first husband and wife team to write and create images together in the field. Their book *Camera Shots at Big Game* has an introduction by Theodore Roosevelt. James Ricalton is adventuring across India and China shooting stereographs for publisher Underwood & Underwood.

1906
The remarkable photos of George Shiras are finally published in *National Geographic*. There is much debate over whether to include images in this long-running journal. Some board members resign at the thought of such a high-quality magazine turning into a 'picture book'.

1907
The earliest practicable colour process, 'autochrome', goes on sale. The fragile glass plates are manufactured by the Lumière brothers in France. Pathé Frères make footage of elephants in India. Swiss Doctor David produces perhaps some of the earliest moving pictures of wildlife in Africa. Stephen Leek films elk migration in Jackson Hole, Wyoming and later uses his films to 'arouse public concern for the elk' over predator species like wolves.

1908
Georg Shulz makes *Natur-Urkunden*, a series of books 'for all who have a heart for nature'. The Oklahoma Natural Mutoscope Company release *The Wolf Hunt*, showing men, including notorious hunter Jack Abernathy, on horseback running down wolves and coyotes. The film is made at the request of Theodore Roosevelt, who had seen Abernathy catch coyotes with his bare hands in Oklahoma but was not believed by folks back east.

1909
Robert Peary takes his cameras all the way to the North Pole – though his claims are disputed – and writes a do-it-yourself publicity manual *The Kodak at the North Pole.* Douglas English launches his 'One Hundred Photographs from Life' series, with a book about mice and a charming title by Reginald Lodge on British birds. Cherry Kearton heads to Kenya on his first African safari with James Clark of the American Museum of Natural History. They meet up with Roosevelt and his hunting entourage. Kearton's footage includes shots of hippos in a lagoon. Cashing in on the media interest, Otis Turner's copycat documentary *Hunting Big Game in Africa* includes a staged scene of a lion being shot on camera, filmed in Chicago with a Roosevelt lookalike actor.

1911
John Hemment makes what may be the first aerial films of wildlife, shooting a flock of wild ducks 'in a panic'. He originally takes up photography to settle disputes in horse races and becomes the father of the 'photo-finish'. Emma Turner's image of a nesting bittern in Norfolk is the first evidence of its return to Britain after its local extinction in the nineteenth century. The Royal Society for the Protection of Birds campaigns to stop the trade in wild bird plumes, postering train stations and walking a mobile exhibition of Arthur Mattingley's egret photographs through London.

1912
Intrepid traveller Helen Messinger Murdoch uses autochrome on her round-the-world tour, one of the first women to undertake such a journey, and she shows the results at her lectures.

1913
John Ernest Williamson takes underwater photos 10 m (30 ft) deep and his efforts are published in *Scientific American* as 'Submarine Photography – A New Art'. David Fairchild's incredible macro-shots of insects and spiders appear in *National Geographic* and later his *Book of Monsters*.

1914
Frank Hurley is filming in Antarctica on Douglas Mawson's expedition, with his images seen widely in *Home of the Blizzard* and shown at public lectures. Hurley's photographs are used in a growing international campaign to stop the trade in penguin oil. Wolfgang Köhler's *Intelligence of Apes* shows footage of chimpanzees problem solving and his later research is a turning point in the psychology of thinking. Williamson and his brother

Frederick Champion, tiger by flashlight, 1925

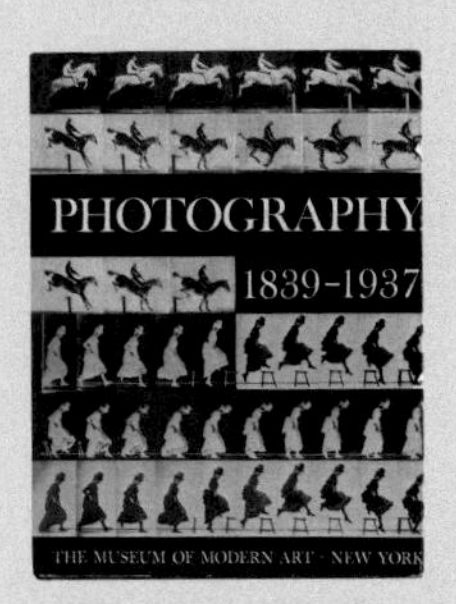

Photography 1839–1937, exhibition catalogue, 1937

Camilla Koffler, *U.S. Camera*, 1940

David Attenborough, *Zoo Quest*, 1961

George make *Terrors of the Deep*, an early underwater picture in the Bahamas, using a dead horse as bait to lure sharks into shot. It is also released as *Thirty Leagues Under the Sea*.

1916
Carl Akeley patents his Akeley camera, finding his Urban bioscope not up to the job in Africa. In 1921, Akeley films gorillas in the Virunga Mountains for the American Museum of Natural History. They are the first photos, still or moving, of these great apes in the wild.

1925
The Leitz company launches the Leica. With fast shutters, light and compact, it is perfect for outdoor shots. Arthur Dugmore travels to Africa with a cine-camera and manages to capture footage at Ngorongoro, including a charging elephant without it being killed. In northern India, Frederick Champion obtains dozens of remarkable night-time photographs, many of which were the first of wild tigers, leopards and sloth bears. His commitment inspires many hunters to lay down their guns, including his friend Jim Corbett with whom he helps found India's first national park in 1935.

1926
William Longley and Charles Martin capture an image of a hogfish, the first underwater creature photographed in colour in the waters off Dry Tortugas, Florida.

1930
35mm cameras like the Leica are now very popular, their speed and compactness making them ideal for outdoor photography. Everything from pets and birdwatching to hunting and skiing. Films in the 'Secrets of Nature' series, like *Bathtime at the Zoo* and *Sweet Peas*, are now using sound and music. Kodak release the 'Kodatoy' 16mm movie projector. A range of films are offered, including 'thrilling jungle adventurers', Charlie Chaplin shorts and cartoons with Felix the Cat or Mickey Mouse.

1931
Ernest Schoedsack shoots *Rango*, a film about an orphan orangutan for Paramount. Like modern wildlife films that follow, it innovates scripted action and voice-over narration. Films created for big Hollywood studios are using large amounts of animal footage, both staged confrontations and natural behaviour.

1935
Arthur Allen, an ornithologist from Cornell University, captures what is thought to be the only film record of the now extinct ivory-billed woodpecker.

1936
LIFE magazine relaunches as the first all-photographic American news magazine and it dominates the market for decades.

1937
MoMA's first photography show, 'Photography 1839–1937' opens, the most comprehensive yet held in America. A handful of animal photos make the cut: Muybridge's motion studies of jumping horses and running deer, X-ray radiographs of fish and snakes by Eder and Valenta and a Rolleiflex shot of a hippo's mouth by Camilla 'Ylla' Koffler.

1940
Ylla's work is featured in *U.S. Camera* magazine. She immigrates to America on a visa sponsored by MoMA. She had started photographing animals in 1932, opening a studio in Paris for pets, and becomes one of the leading animal photographers of her generation.

1950
Televisions make their way into homes across the world. Photographs and film are viewed as never before. Organizations are starting to use moving footage in their efforts to protect animals.

1951
Jacques Cousteau begins his career as a broadcaster, outfitting the ship *Calypso* with funds from the French Navy and the Ministry of Education. Hans Hass wins first prize at Venice Film Festival for feature-length documentary *Under the Red Sea*. A Disney 'True Life Adventure' short-reel, *Nature's Half Acre*, wins an Academy Award.

1954
David Attenborough heads to Sierra Leone with zoologists Jack Lester and Alfred Woods and cameraman Charles Lagus to film them collecting animals for London Zoo. His first on-screen appearance is 21 December, introducing *Zoo Quest*. A second series takes him to Borneo in search of the Komodo dragon.

1957
Russell Kirsch produces the first digital photograph, after scanning a film photograph of his son into a computer. A small square image, it has a resolution of 176 × 176 pixels. *LIFE* magazine later declares it to be one of the '100 photographs that changed the world'. It's another twenty years before Kodak engineer the first digital camera and not until the 1990s that digital image quality really begins to compete with film.

1958
Disney's Arctic adventure *White Wilderness*, with its now-notorious faked lemming

Comic book for the film *White Wilderness*, 1958

Jacques Cousteau's 'diving saucer', 1959

Armand Denis's first issue of *Animals* magazine, 1963

Jane Goodall and Hugo van Lawick observe chimpanzees, 1965

sequence, wins the Academy Award for Best Documentary Feature.

1960
Cousteau continues to document the undersea realm on film. With Jean de Wouters, he has been developing the first submersible 'amphibious' 35mm camera. His waterproof design is bought by Nikon in Japan and released widely as the 'Nikonos' in 1963.

1963
Belgian film-maker Armand Denis launches *Animals* magazine, championing the best in wildlife photography. Two decades later, the title becomes *BBC Wildlife*. Early issues carry extracts from Rachel Carson's important book *Silent Spring*, which details for the first time the horrors of the pesticide industry and its impact on wildlife. Jane Goodall's work with chimpanzees in Tanzania is featured in *National Geographic*. A shot of Goodall observing the chimpanzees is later selected by NASA to be encoded in analog on a golden phonograph record and launched into space on the Voyager probe in 1977.

1965
Animals establishes the Wildlife Photographer of the Year competition. It attracts over three hundred entries in its first year; David Attenborough is on hand to congratulate winner Roger Dowdeswell, whose photo is a colour portrait of a tawny owl bringing prey for its chicks.

1967
The World About Us premieres on BBC2, the first colour broadcasts of wildlife films in Britain. The first film shows the private life of kingfishers. In New York, Garry Winogrand is shooting in zoos in Central Park and the Bronx. He eventually publishes four books, including *The Animals* in 1969. Winogrand uses a small-format, 35mm camera that enables him to photograph quickly and freely. At the time of his death in 1984, he left more than 2,500 undeveloped rolls of film.

1972
Bill Mason's film *Cry of the Wild* is released in theatres across North America and earns millions at the box-office. His work does much to change the reputation of wolves, who are 'more victims than killers'. Cousteau is working on *Ocean World*, a huge marine encyclopaedia series filled with photos. It is released over the following years in twenty-one volumes.

1975
Masahisa Fukase begins his series *Ravens* with a chance photograph on his native island Hokkaido, Japan. For almost a decade he documents the birds obsessively, creating haunting and melancholic images, an elegy of solitude and loss.

1979
Attenborough releases the landmark thirteen-part series *Life on Earth*. It is a huge international success and is followed by *The Living Planet* in 1984 and *The Trials of Life* in 1990.

1980
George Schaller becomes the first outsider ever permitted by the Chinese government to study pandas in the wild, joining a team of field researchers in the mountains of Sichuan. The initiative is enabled by conservation charity the WWF (World Wildlife Fund for Nature), who has used a stylized panda as its logo since it was founded in 1961. Schaller's insights and photographs are published in *National Geographic* in December 1981 and his later book *The Last Panda* is a bestseller.

1981
Canon debuts the first of its adverts 'Wildlife As Canon Sees It' in *National Geographic*: each month a full page telling the story of a different endangered species. By the final instalment in December 2020, some 466 species have been featured.

1989
The Internet makes millions of pictures available at the click of a button.

1992
The first email attachment is sent: a photo of a barbershop quartet. Researcher Nathaniel Borenstein invented the extension code to enable it, as he dreams one day of being able to see pictures of his grandchildren over email.

2000
When Sharp launch the world's first camera phone, the Sharp J-SH04, Japanese teenagers buy them in their millions. Samsung release the SCH-V2000 with a built-in camera capable of taking twenty photos with 0.35 mega-pixel resolution. In the US, Sanyo release a chunky clamshell design with flash and basic digital zoom.

2003
Camera phones are really taking off with over 80 million already sold worldwide. The first video ever posted on YouTube is a short clip of co-founder Jawed Karim checking out the elephants at the San Diego Zoo.

2007
Netflix delivers its billionth DVD, but moves rapidly towards a new business model: streaming media and video on demand. In 2019, Netflix releases its first nature documentary series *Our Planet*, narrated by Attenborough. By 2021, it is reported that over 100 million households have watched the show.

Richard Vevers captures the first underwater images for Google Maps, 2012

DJI's Phantom 4 quadcopter, 2016

Relicanthus sp., a new species discovered in the Clarion-Clipperton Zone, 2016

2010
Though Apple's first iPhone arrived in 2007, it is the iPhone 4 that proves a game-changer: the first to have a front-facing camera. Selfies take over the world. Smartphones and tablets are being used to create and share millions of images. Before it was even called Instagram, founder Kevin Systrom uploads the first photo, taken near a taco stand in Mexico. The caption that accompanies it: 'test'. No surprise, it's a photo of a dog.

2012
Search giant Google adds the first underwater panoramic images to Google Maps so 'anyone can become the next virtual Jacques Cousteau'. Google partners with Catlin Seaview Survey, a major scientific study of the world's coral reefs.

2013
Drone company DJI launch the Phantom. The first camera-equipped drones enter the commercial market. 360Heroes release the scuba edition of its system for capturing 360-degree video. Drones soon become a viable tool for film-makers and for studying wildlife populations. Professional octocopters take cameras into previously hard-to-reach areas, at considerably less cost and risk than small aircraft or helicopters. *National Geographic* celebrates its 125th birthday with a special photography issue. Editor-in-chief Susan Goldberg declares: 'No magazine has taken its readers more places – into the human past, into the minds of animals, into the deepest parts of the oceans and the ends of the known universe. Our photographers, writers, artists, and editors will keep telling those stories at the same time we investigate our global future'.

2016
In the Pacific Ocean's Clarion-Clipperton Zone, a 5,000-m (18,000-ft) abyssal plain off Hawaii, remote photography brings to light species previously unknown to science.

2018
Images taken from orbiting satellites help scientists understand more about animal populations: new penguin colonies are discovered due to guano stains visible from space, whilst high-resolution images of whales enable researchers to learn more about whale abundance in the most remote places.

2021
The search for new life continues. A robo-drone named Ingenuity, weighing 1.8 kg (4 lb), is used to investigate if there has ever been life on Mars.

TODAY
It is estimated 4.7 billion photographs are created every day – 54,400 per second. That's 1.72 trillion photos created worldwide in a year. Sophisticated digital cameras and social-media platforms give all kinds of photographers the chance to use their imagery in support of animals and conservation efforts. Most people are happy enough just to create pictures of their pets. It has been calculated that over 3 billion of those images, and over 720,000 hours of video, are shared online each day. That's a whole lot of cats and dogs.

(*right*) Tim Flach, panda control room at the National Zoo, Washington, for *Smithsonian* magazine, 2014
(*opposite*) Bob Wallace, Lotus the Nile hippo, California, 1937

Further Reading

FROM THE INVENTION of the camera to this bewildering digital age, animals have regularly been the focus of attention. There has been a staggering evolution of technologies and techniques, and photographers continue to extend what is photographically possible. Though animals are everywhere in books, it's surprising that wildlife photography as a genre is rarely included in conventional photographic histories, and until only recently has animal photography been regarded as a fine art.

This book is intended as an extended visual essay on the history and future of animal photography. It's a sequence of images and insights, gathered to encourage us to think about the ways we engage with animals, and to reconsider our relationship with nature. I've researched images to better understand *why* people have photographed animals at different points in the history of the medium and to explore what photographing animals means in a rapidly changing world.

This is deliberately not a conventional account of wildlife photography. There are annual collections of competition images if that is your thing. Rather, this book is an exploration of the animal *in* photography. It speaks to our ongoing desire to look at animals – to use and abuse them, just as we appreciate and protect them – and photography's role in all this. It's a story of an awakening of animal awareness through the visual image: a story of creativity, yes, but also of crisis.

As a medium driven by technology and creativity, photography is constantly on the move and its social functions are changing. I've tried to bring forward a combination of famous and important work, as well as little-known treasures from the archives. For more insights and inspiration, here are thirty other books that will send you off in new directions:

Robert Adams, *Why People Photograph* (New York: Aperture, 2004)

Yann Arthus-Bertrand and Brian Skerry, *From Above and Below* (London: Thames & Hudson, 2013)

Gerry Badger, *The Genius of Photography* (London: Quadrille, 2014)

Jean-Christophe Bailly, *George Shiras* (Paris: Xavier Barral, 2015)

Steve Baker, *Picturing the Beast* (Urbana: University of Illinois Press, 2001)

James Balog, *Survivors* (New York: Abrams, 1990)

John Berger, *About Looking* (New York: Pantheon, 1980)

Richard Bernabe, *Wildlife Photography* (London: Ilex, 2018)

John Bevis, *The Keartons* (Axminster: Uniformbooks, 2016)

Derek Bousé, *Wildlife Films* (Philadelphia: University of Pennsylvania Press, 2000)

John Bradshaw, *The Animals Among Us* (London: Allen Lane, 2017)

Nick Brandt, *Inherit the Dust* (New York: DAP, 2016)

Matthew Brower, *Developing Animals* (Minneapolis: University of Minnesota Press, 2011)

Jonathan Burt, *Animals in Film* (London: Reaktion, 2002)

Henry Carroll, *Animals* (New York: Abrams, 2021)

Graham Clarke, *The Photograph* (Oxford University Press, 1997)

Mark Cousins, *The Story of Looking* (Edinburgh: Canongate, 2017)

Margo DeMello, *Animals and Society* (New York: Columbia University Press, 2012)

David Doubilet, *Water Light Time* (London: Phaidon, 1999)

Tim Flach, *Endangered* (New York: Abrams, 2017)

Jean-Marie Ghislain, *Shark* (London: Thames & Hudson, 2014)

Stephen Gill, *The Pillar* (Skåne: Nobody, 2019)

Charles Guggisberg, *Early Wildlife Photographers* (Exeter: David & Charles, 1977)

Donna Haraway, *When Species Meet* (Minneapolis: University of Minnesota Press, 2007)

Melissa Harris, *A Wild Life* (New York: Aperture, 2017)

Marvin Heiferman, *Photography Changes Everything* (New York: Aperture, 2012)

Hal Herzog, *Some We Love, Some We Hate, Some We Eat* (London: HarperCollins, 2010)

Eric Hosking, *Classic Birds* (London: HarperCollins, 1993)

Rosamund Kidman Cox, *The Masters of Nature Photography* (London: Natural History Museum, 2013)

Frans Lanting, *Into Africa* (London: Earth Aware, 2017)

Philip Lymbery, *Dead Zone* (London: Bloomsbury, 2018)

John Mitchell, *The Wildlife Photographs* (Washington: National Geographic, 2001)

Gregg Mitman, *Reel Nature* (Seattle: University of Washington Press, 1999)

Paul Nicklen, *Born to Ice* (Kempen: teNeues, 2018)

Gemma Padley, *Into the Wild* (London: Laurence King, 2021)

Richard Peters, *Wildlife Photography at Home* (London: Ilex, 2019)

Peter Pickford and Beverly Pickford, *Wild Land* (London: Thames & Hudson, 2018)

Nigel Rothfels, *Representing Animals* (Bloomington: Indiana University Press, 2002)

Sebastião Salgado, *Genesis* (Cologne: Taschen, 2013)

Joel Sartore, *The Photo Ark* (Washington: National Geographic, 2017)

Stephen Shore, *The Nature of Photographs* (London: Phaidon, 2010)

Kelley Wilder, *Photography and Science* (London: Reaktion, 2008)

Art Wolfe, *The New Art of Photographing Nature* (New York: Amphoto, 2013)

I said thirty books, sorry, that's forty-three! It's always hard to choose but with wide-ranging projects like this it's almost impossible: there is just so much that could be included. Another way into this field would be to search out each of our contributing photographers. Their social media feeds are listed overleaf. Many of them have published monographs and regularly contribute to print media across all kinds of contemporary visual and storytelling platforms.

So, happy foraging. Keep your eyes and your mind wide open. Always take a camera with you, of whatever kind works best for your purpose. You'll need field-skills, patience, courage, tenacity, quiet determination and imagination, as much as a good waterproof jacket and a thermos for your tea. Make images that matter to you and take your work seriously, however playful or powerful it might be. 'We often talk of saving the planet,' says David Attenborough, 'but the truth is that we must do these things to save ourselves. With or without us, the wild will return.' Or, as Alec Soth put it: 'If in your heart of hearts you want to take pictures of kitties, take pictures of kitties.'

Contributors

Ingo Arndt
W: ingoarndt.com

Levon Biss
IG: @levonbiss

Xavi Bou
IG: @xavibou

John Bozinov
IG: @johnbozinov

Will Burrard-Lucas
IG: @willbl

Stefan Christmann
IG: @christmannphoto

Tim Flach
IG: @timflachphotography

Daisy Gilardini
IG: @daisygilardini

Sergey Gorshkov
IG: @sergey_gorshkov_photographer

Melissa Groo
IG: @melissagroo

Karim Iliya
IG: @karimiliya

Britta Jaschinski
IG: britta.jaschinski.photography

Leila Jeffreys
IG: @leilajeffreys

Kate Kirkwood
IG: @k8kirkwood

Tim Laman
IG: @timlaman

Florian Ledoux
IG: @florian_ledoux_photographer

Dina Litovsky
IG: @dina_litovsky

Jo-Anne McArthur
IG: @weanimals

Daniel Naudé
IG: @daniel_naude

Jim Naughten
IG: @jimnaughten

Anuar Patjane
IG: @anuarpatjane

Mateusz Piesiak
IG: @mpwildlife

Alicia Rius
IG: @aliciariusphoto

Claire Rosen
IG: @claire__rosen

Marcin Ryczek
IG: @marcinryczek_photography

Traer Scott
IG: @traer_scott

Alexander Semenov
IG: @narwhal_season

Anup Shah
W: anupshah.com

Nichole Sobecki
IG: @nicholesobecki

Paul Souders
IG: @paul.souders

Georgina Steytler
IG: @georgina_steytler

Marsel van Oosten
IG: @marselvanoosten

Staffan Widstrand
IG: @staffanwidstrand

Shannon Wild
IG: @shannon__wild

Steve Winter
IG: @stevewinterphoto

Kiliii Yuyan
IG: @kiliiiyuyan

Sources of Quotations

Page 7 'I like dogs...', Elliott Erwitt, as quoted in *Dogs, Dogs* (London: Phaidon, 1998); 'Photography's ubiquity', Jeffrey Fraenkel, *Long Story Short* (San Francisco: Fraenkel Gallery, 2019); 'Nature is painting...', John Ruskin, *The True and the Beautiful in Nature* (New York: John Wiley, 1860). **Page 10** 'The truth is...', Robert Capa, interview, *World-Telegram*, 2 September 1937; 'What's happening...', John Szarkowski as quoted in *The Photographer's Eye* (New York: The Museum of Modern Art, 2007); 'Beauty can be...', Richard Misrach, interview, *Violent Legacies* (Manchester: Cornerhouse, 1992). **Page 12** 'No one will protect...', David Attenborough, interview, *Ecologist*, 4 April 2013. **Page 17** 'When we contemplate...', John Muir, *Travels in Alaska* (Boston: Houghton Mifflin, 1915). **Page 18** 'No human hand...', Michael Faraday, 'The New Art', *The Literary Gazette* (2 February 1839). **Page 26** 'At this moment', William Bradford, *The Arctic Regions* (London: Sampson Low, 1873). **Page 28** 'Now every nipper...', Alvin Langdon Coburn, 'The Future of Pictorial Photography', *Photograms of the Year* (London: Hazell, Watson & Viney, 1916). **Page 29** 'The photographer is like...', George Bernard Shaw, quoted in Helmut Gernsheim, *A Concise History of Photography* (London: Thames & Hudson, 1965). **Page 30** 'The small attempts...', Elizabeth Eastlake, 'Photography', *Quarterly Review* (April 1857). **Page 106** 'Some we love...', Hal Herzog, *Why It's So Hard to Think Straight About Animals* (London: HarperCollins, 2010). **Page 119** 'In all things...', Aristotle, *De Partibus Animalium*, trans. James Lennox, *On the Parts of Animals* (Oxford: Clarendon, 2002); 'Cavalcade of animals...', Marcel Ravidat, quoted in *History Today*, 9 September 2015; 'It was a challenging...', Ralph Morse, 'Cave Paintings', *LIFE*, 24 February 1947. **Page 122** 'I wanted to show...', Nina Leen, 'Discovering the Beauty of Bats', *LIFE*, 29 March 1968. **Page 123** 'Knowing does not...', Sylvia Earle, *The World is Blue* (Washington: National Geographic, 2009); 'I'm not an animal lover...', David Attenborough, interview, *Metro*, 29 January 2013. **Page 124** 'People must feel...', David Attenborough, *A Life on our Planet* (London: Ebury, 2020). **Page 125** 'It felt like...', Ami Vitale, interview, *National Geographic*, October 2019. **Page 126** 'That image and story...', Paul Nicklen, 'N.G. Live!', *National Geographic*, March 2014. **Page 315** 'I do not...', William Henry Fox Talbot, 'The New Art', *The Literary Gazette* (2 February 1839); 'The best arguments...', Richard Powers, *The Overstory* (New York: Random House, 2019); 'Artists must confront...', Ben Okri, 'Opinion', *The Guardian*, 12 November 2021; 'Zoos, realistic animal toys...', John Berger, 'Why Look at Animals', *About Looking* (New York: Pantheon, 1980); 'After all, what is...', William Henry Fox Talbot, letter to the editor of the *Literary Gazette*, reproduced in 'The New Art', *Literary Gazette* (2 February 1839) and 'The Pencil of Nature', *Blackwood's Edinburgh Magazine* (March 1839) and *Corsair* (13 April 1839). **Page 316** 'Because I like...', Jean-Louis Etienne, in Huw Lewis-Jones and Kari Herbert, *In Search of the South Pole* (London: Conway, 2011); 'Generation Dread'..., Britt Wray, *Finding Purpose in an Age of Climate Crisis* (Toronto: Knopf, 2022); 'The best antidote...', Chase Jarvis, *Creative Calling* (New York: HarperBusiness, 2019); 'Instructions for living...', Mary Oliver, 'Sometimes', *Red Bird* (Boston: Beacon Press, 2008). **Page 317** 'Most animals...', and 'This is my life...', interviews with the author. **Page 329** 'We often talk...', David Attenborough, *A Life on our Planet* (London: Ebury, 2020); 'If in your heart...', Alec Soth, first mentioned in a Magnum Photos blog post, also *Wear Good Shoes* (Paris: Magnum Photos, 2019).

Illustration Credits

a=above, b=below, c=centre, l=left, r=right

Front endpapers, row 1 The J. Paul Getty Museum, Los Angeles; Library of Congress Prints & Photographs Division, Washington D. C.; The J. Paul Getty Museum, Los Angeles; National Galleries of Scotland; **row 2** National Galleries of Scotland; Smithsonian Institution, Washington D. C.; The J. Paul Getty Museum, Los Angeles; **row 3** The J. Paul Getty Museum, Los Angeles; Library of Congress Prints & Photographs Division, Washington D. C.; Library of Congress Prints & Photographs Division, Washington D. C.; The J. Paul Getty Museum, Los Angeles; **row 4** New York Public Library; National Galleries of Scotland; New York Public Library; **opposite title page, row 1** National Galleries of Scotland; The J. Paul Getty Museum, Los Angeles; The J. Paul Getty Museum, Los Angeles; **row 2** Library of Congress Prints & Photographs Division, Washington D. C.; **row 3** The J. Paul Getty Museum, Los Angeles; New York Public Library; **row 4** New York Public Library; Library of Congress Prints & Photographs Division, Washington D. C.; **2–3** © Daisy Gilardini; **4** Lay Auctioneers/BNPS; **6** Private Collection; **8** The J. Paul Getty Museum, Los Angeles; **9** © Daniel Szalai; **10** Roxana Dulama/Caters News; **11** Ellis Collection of Kodakiana/Duke University Library; **12** John Downer Productions; **13** V&A Images/Getty Images; **14–15** © Adam Oswell/We Animals Media; **16** Metropolitan Museum of Art, New York; **18** The Picture Art Collection/Alamy Stock Photo; **19** Metropolitan Museum of Art, New York; **20al** The Nelson-Atkins Museum of Art, Kansas City, Missouri. Gift of the Hall Family Foundation, 2010.72.2. Image courtesy Nelson-Atkins Media Services; **20ar** Private collection; **20bl** The Nelson-Atkins Museum of Art, Kansas City, Missouri. Gift of the Hall Family Foundation, 2005.37.10. Image courtesy Nelson-Atkins Media Services; **21a** The Nelson-Atkins Museum of Art, Kansas City, Missouri. Gift of Hallmark Cards, Inc., 2005.27.37. Image courtesy Nelson-Atkins Media Services; **21bl** The J. Paul Getty Museum, Los Angeles; 21br The Nelson-Atkins Museum of Art, Kansas City, Missouri. Gift of Hallmark Cards, Inc., 2005.27.77. Image courtesy Nelson-Atkins Media Services; **22a** The Picture Art Collection/Alamy Stock Photo; **22b** Metropolitan Museum of Art, New York; **23** National Museum Wales; **24a** Royal Collection Trust/© His Majesty King Charles III 2023; **24b & 25a** The J. Paul Getty Museum, Los Angeles; **25b** Reginald Lodge; **26** British Library, London; **27** Mary Evans Picture Library; **28** National Science and Media Museum, Bradford; **29a** Private Collection; **29b** World History Archive/Alamy Stock Photo; **30** Metropolitan Museum of Art, New York; **31a** Martin and Osa Johnson Safari Museum, Chanute, KS; **31b** Charles Martin; **33** State Library of New South Wales, Sydney; **34–35** © Zoological Society of London/Bridgeman Images; **37, 38–39, 40–41** © Florian Ledoux; **43, 44, 45, 46–47** © Ingo Arndt; **49, 50, 51** © Traer Scott; **53, 54, 55, 56–57** © Kate Kirkwood; **59, 60, 61, 62–63** © Jo-Anne McArthur; **64, 66, 68–69, 70, 71–72** © Jim Naughten; **75, 76, 77, 78, 79** © Daniel Naudé; **81, 82–83, 84–85** © Georgina Steytler; **87, 88–89, 90–91** © Mateusz Piesiak, www.mateuszpiesiak.pl; **93, 94, 95, 96–97** © Karim Iliya; **98, 100, 101, 102–103, 105** © Britta Jaschinski; **107** Arcaid Images/Alamy Stock Photo; **108** The J. Paul Getty Museum, Los Angeles; **109** Picture Post, August, 1949; **110a** David Fairchild; **110b** © Thomas Marent/naturepl.com; **111** eye of science; **112a** © Bruno D'Amicis/naturepl.com; **112b** Fox Photos/Hulton Archive/Getty Images; **113** imageBROKER.com GmbH & Co. KG/Alamy Stock Photo; **114a** Metropolitan Museum of Art, New York; **114b** The J. Paul Getty Museum, Los Angeles; **115a** Metropolitan Museum of Art, New York; **115b** © Tony Wu/naturepl.com; **116–117** © James Mollison; **118** Science History Images/Alamy Stock Photo; **120** Ralph Morse/The LIFE Picture Collection/Shutterstock; **121a** Fritz Goro/The LIFE Picture Collection/Shutterstock; **121b, 122** Nina Leem/The LIFE Picture Collection/Shutterstock; **123** PA Images/Alamy Stock Photo; **124** National Geographic; **125** © Ami Vitale; **127** Brett Stirton/Getty Images; **128–129** © Charles Hamilton James/naturepl.com; **131, 132, 133, 134–135** © Xavi Bou; **137, 138, 139, 140–141** © Alexander Semenov; **143, 144–145, 146–147** © Sergey Gorshkov; **149, 150, 151** © Tim Laman; **153, 154, 155** © Melissa Groo; **156, 159, 160–161, 163, 164–165** © Dina Litovsky; **166, 168, 170, 171, 173, 174–175** Photo by Tim Flach; **177, 178, 179** © Stefan Christmann/Nature in Focus; **181, 182, 183, 184, 185** © Leila Jeffreys; **187, 188–189, 190–191** © Anuar Patjane Floriuk; **193, 194, 195, 196, 197** © John Bozinov; **198, 200, 202–203, 204** © Claire Rosen; **206** © Evgenia Arbugaeva; **207** © Morgan Heim; **208** Tasmanian Museum and Art Gallery; **209a** © Maroesjka Lavigne, courtesy of Robert Mann Gallery, New York; **209b** Burton Historical Collection, Detroit Public Library; **210a** New York Public Library; **210b** © Mitsuaki Iwago/naturepl.com; **211a** © Andreas Gursky/Courtesy Sprüth Magers Berlin London/DACS 2023; **211b** New York Public Library; **212** Photograph George Shiras; **213** Ed Ram/Getty Images; **214a** V&A Images; **214b** © Chris Jordan; **215** © Daniel Beltrá; **216** © Brian Skerry; **217** © Margaret Bourke-White/The LIFE Collection/Shutterstock; **219, 220–221, 222–223** © Marcin Ryczek; **225, 226, 227, 228–229** Nichole Sobecki/VII; **231, 232–233, 234–235** © Marsel van Oosten; **237, 238, 239, 240–241** Staffan Widstrand/Wild Wonders International; **243, 244–245, 246-247** © Alicia Rius; **248, 251, 252–253, 254** © Levon Biss; **256, 258–259, 260, 262–263** © Steve Winter; **265, 266, 267, 268–269** © Shannon Wild; **271, 272, 273, 274–275** © Kiliii Yuyan; **277, 278–279, 280, 281** © Daisy Gilardini; **283, 284, 285, 286–287** © Anup Shah/www.shahrogersphotography.com; **289, 290–291, 292–293** © Will Burrard-Lucas; **294, 297, 298–299, 301** © Paul Souders/Worldfoto; **302** © Martin Parr/Magnum Photos; **303** The Eric Hosking Charitable Trust; **304a** Private Collection; **304b** © Richard Barnes; **305a** © David Chancellor; **305b** The J. Paul Getty Museum, Los Angeles; **306-307** © Werner Bischof/Magnum Photos; **308** © Elliott Erwitt/Magnum Photos; **309a** Matt Stuart - Trafalgar Square 2004; **309b** © Richard Peters; **310a** © Fondation Henri Cartier-Bresson/Magnum Photos; **310b** The J. Paul Getty Museum, Los Angeles; **311** © Stephen Gill; **312–313** © Dmitry Kokh; **314** SSPL/Getty Images; **316** Private Collection; **317** © Marsel van Oosten; **318–319** © Sergey Gorshkov; **320** Rijksmuseum, Amsterdam; **321a** Cleveland Museum of Art; **321ca** Metropolitan Museum of Art, New York; **321cb** National Museum Wales; **321b** The Picture Art Collection/Alamy Stock Photo; **322a** Metropolitan Museum of Art, New York; **322ca** The J. Paul Getty Museum, Los Angeles; **322c** Bettmann/Getty Images; **322cb** Metropolitan Museum of Art, New York; **322b** Private Collection; **323a** Bettmann/Getty Images; **323ca & cb** Private Collection; **323b** Library of Congress Prints & Photographs Division; **324a** Frederick Champion; **324ca** MoMA, New York; **324cb, b & 325a** Private Collection; **325ca** Jack Garofalo/Paris Match/Getty Images; **325cb & b** Private Collection; **326a** Underwater Earth/XL Catlin Seaview Survey/Christophe Bailhache; **326ca** A. Savin, WikiCommons; **326cb** Image courtesy of Craig Smith and Diva Amon, ABYSSLINE Project; **326b** Photo by Tim Flach; **327** George Wallace; **back endpapers** © Jim Naughten.

Author Biography

HUW LEWIS-JONES is an award-winning author, educator, and expedition leader with a PhD from the University of Cambridge. He is now a professor at Falmouth University, where he encourages his natural history students to explore the multiple meanings of animals. Published in twenty languages, his books include *The Sea Journal*, *Explorers' Sketchbooks*, *Imagining the Arctic*, *Swallowed by a Whale*, and *The Writer's Map*, a bestselling atlas of imaginary lands. He is also the author of many books for children including *Blue Badger*, *Clive Penguin*, and the *Bad Apple* series. When not teaching, Huw escapes into wilderness areas and has been twelve times to the North Pole.

IG: @huwlewisjones

Acknowledgments

THIS BOOK WOULD be nothing without the animals. Sure, our team of photographers has trekked through jungles, up mountains, across deserts and ice-floes, enjoying storm, snow and sunshine to bring back their remarkable images, but it's the animals that we must really celebrate. At a time when more and more people are aware of the wonder of nature, through photography, it seems a curious kind of madness that animals are suffering as never before.

Over my career as a wilderness guide and naturalist, I've been lucky enough to navigate little boats to remote islands, through Arctic ice, among reefs in the Pacific and along some of Antarctica's most desolate coasts, and all the while in appreciative awe of the wildlife I've encountered. But now I begin to wonder: are we part of the problem? Do not expeditions and nature tourism, however sustainably and responsibly practised, risk everything that we love? Is photography itself – in drawing more people onto aeroplanes and out into precious habitats – contributing to the ruining of the great outdoors? These are worthy questions that my students and I grapple with each week. Can photography be a force for good in this world? I do hope so.

I'm grateful to all the photographers who I've interviewed for this book and who kindly lent their insights and imagery. I'm especially thankful to my companions on many voyages, in particular John Bozinov and artist Spider Anderson, who have encouraged me to think in new ways. And veteran Icelandic photojournalist Ragnar Axelsson, whom I've had the pleasure of knowing for over a decade, since our first project *Last Days of the Arctic*. More recently, I've been on ships with peerless photographers Daisy Gilardini and Sergey Gorshkov, who you can find in the book, and Camille Seaman and James Balog, who aren't but whose wisdom has shaped it in different ways. It was a particular pleasure for me to give Leila Jeffreys her first taste of the Arctic a few years ago, and I look forward to searching out more seabirds with her.

At Thames & Hudson, my new editor Emma Barton rose to the challenge of helping me wrestle the book into shape, while Andrew Sanigar led the way. Thanks also to designer Daniele Roa and Sally Nicholls, who followed my research trail through international archives. I appreciate all their talents hugely. From the edge of the Pacific, Yosef Wosk once more supported this project with generosity and wisdom. And lastly, to my marine and natural history students here in Cornwall. Yours is the future. I'm sorry that our natural world is being handed to you in such a state. Keep using your skills to stand up and speak for those you care about. And remember, outside is always better than online.

Index

Page numbers in *italics* refer to illustrations.

To my MNHP students – I hope you find this useful!

FRONT COVER: Tim Flach, *Windows Chestnut*, from the series *Equus*, 2008. © Tim Flach.

BACK COVER (clockwise from top left): Anonymous daguerreotype, 1850s. Photograph Sotheby's/Private Collection; Paul Souders, polar bear, Hudson Bay, Canada, 2013. © Paul Souders/Worldfoto; Alexander Semenov, lion's mane jellyfish, in the strait of Velikaya Salma, White Sea, 2019. © Alexander Semenov; Leila Jeffreys, Major Mitchell's cockatoo, 2012. © Leila Jeffreys.

FRONT ENDPAPERS: The first mass-produced animal images, stereographs from international collections, 1865–1904.

BACK ENDPAPERS: Jim Naughten, stereographs from the series *Mountains of Kong*, 2019.

PAGE 2: Daisy Gilardini, Weddell seal, Half Moon Island, Antarctica, 2014.

PAGE 4: Frederick York, hand-painted magic lantern slides, London Zoo, 1870.

First published in the United States of America in 2024 by Thames & Hudson Inc., 500 Fifth Avenue, New York, New York 10110

Library of Congress Control Number 2023943228

ISBN 978-0-500-02272-6

Printed and bound in China by Toppan Leefung Printing Limited

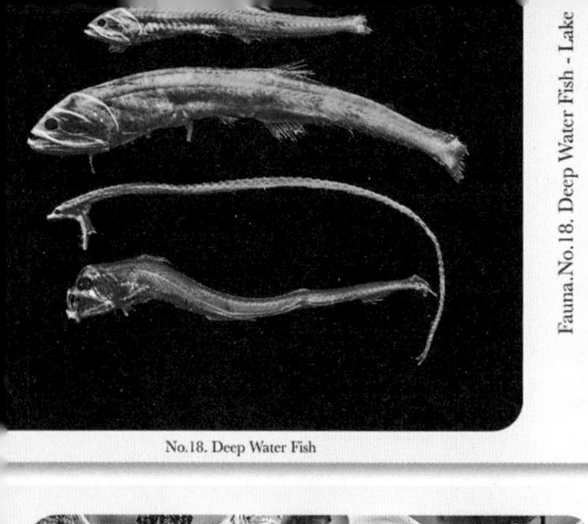

Fauna.No.18. Deep Water Fish - Lake

No.18. Deep Water Fish

Jim Naughten Presents 'Mountains of

Stereoscopic images from a lost landscape

No.09.The

Mountains of Kong

Fauna.No.13. The Insects

No.13. The Insects

Jim Naughten Presents 'Mountains of Kong'

Stereoscopic images from a lost landscape

Mountains of Kong

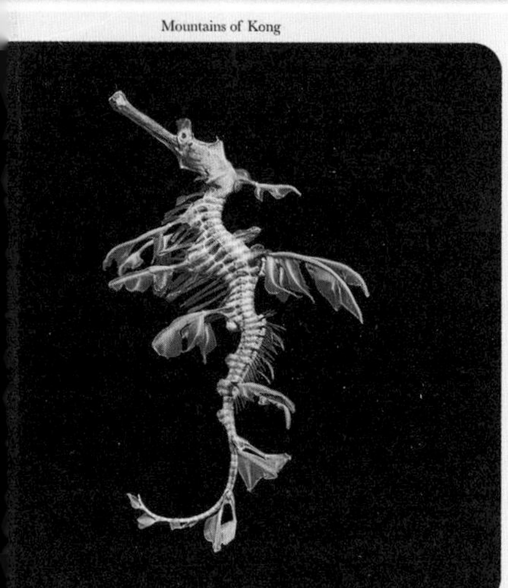

Fauna.No.11. The Sea Dragon

No.11. The Sea Dragon

Jim Naughten Presents 'Mountains of Kong'

Stereoscopic images from a lost landscape

Mountains of Ko

No.15. The Anima

Mountains of Kong

.No.02. The Monkey Tree

n Presents 'Mountains of Kong'

ic images from a lost landscape

Fauna.No.01. The Lions
Jim Naughten Presents 'Mountains of
Stereoscopic images from a lost landscape
No.10.The Jungle Gorilla

Mountains of Kong
Fauna.No.06. The Silver Lion
No.06. The Silver Lion
Jim Naughten Presents 'Mountains of Kong'
Stereoscopic images from a lost landscape

Fauna.No.15. The Animal Pass
Jim Naughten Presents 'Mountains of Kong'
Stereoscopic images from a lost landscape
Mountains of Kong
No.04.The Deer

Mountains of Kong